About Island Press

Since 1984, the nonprofit organization Island Press has been stimulating, shaping, and communicating ideas that are essential for solving environmental problems worldwide. With more than 1,000 titles in print and some 30 new releases each year, we are the nation's leading publisher on environmental issues. We identify innovative thinkers and emerging trends in the environmental field. We work with world-renowned experts and authors to develop cross-disciplinary solutions to environmental challenges.

Island Press designs and executes educational campaigns, in conjunction with our authors, to communicate their critical messages in print, in person, and online using the latest technologies, innovative programs, and the media. Our goal is to reach targeted audiences—scientists, policy makers, environmental advocates, urban planners, the media, and concerned citizens—with information that can be used to create the framework for long-term ecological health and human well-being.

Island Press gratefully acknowledges major support from The Bobolink Foundation, Caldera Foundation, The Curtis and Edith Munson Foundation, The Forrest C. and Frances H. Lattner Foundation, The JPB Foundation, The Kresge Foundation, The Summit Charitable Foundation, Inc., and many other generous organizations and individuals.

The opinions expressed in this book are those of the author(s) and do not necessarily reflect the views of our supporters.

Praise for *Gray to Green Communities*

"Dana Bourland has paved a path for the United States to improve the quality of housing for all—particularly for those whose security and stability depend on affordable housing. A leader in green affordable housing, she has written a must-have book for those seeking to better understand some of the issues facing housing and the green way forward."

—Chrissa Pagitsas, VP, Environmental Social Governance, Fannie Mae

"This crisp book brims with wisdom. Nominally about housing and climate change, it also tackles health risks, poverty, racism, social injustice, and the environmental crisis—and proposes practical, cross-cutting solutions that solve many problems at once. Skillfully deploying both human stories and rigorous data, Dana Bourland gives us a compelling blueprint for sustainable, healthy homes and communities."

—Howard Frumkin, former Director, National Center for Environmental Health, US Centers for Disease Control and Prevention; Professor Emeritus, University of Washington School of Public Health; coeditor of *Making Healthy Places*

"In the wealthiest society in history, no one should live in substandard housing that wastes lives, energy, and resources. *Gray to Green Communities* is a thoughtful, articulate, and impassioned call to do better. Dana Bourland makes a powerful case for addressing the housing crisis systematically, with the tools of integrated design, to help solve problems of climate change, unemployment, and social decay while building cohesive, just, and prosperous communities. A must-read!"

—David W. Orr, author of *The Nature of Design, Earth in Mind,* and coeditor of *Democracy Unchained*

Gray to Green Communities

Gray to Green Communities

A Call to Action on the Housing and Climate Crises

Dana L. Bourland

 **ISLAND**PRESS | Washington | Covelo

Library of Congress Control Number: 2020942997

All Island Press books are printed on environmentally responsible materials.

Manufactured in the United States of America
10 9 8 7 6 5 4 3 2 1

Keywords: Affordable housing; attainable housing; building materials; climate action; climate change; climate justice; climate mitigation; Denny Park, Seattle, Washington; green chemistry; energy burden; energy efficiency; Enterprise Community Partners; environmental justice; green affordable housing; green building; Green Communities Criteria; Green Communities Initiative; green housing; healthy housing; High Point, Seattle; Holland Apartments, Danville, Illinois; housing discrimination; housing justice; integrative design process; Leadership in Energy and Environmental Design (LEED); Living Building Challenge (LBC); Low Income Housing Institute (LIHI); Plaza Apartments, San Francisco, California; resilience; Sanctuary Place, Chicago, Illinois; SeaGreen; sustainable development; United Nations Sustainable Development Goals; US Green Building Council (USGBC)

To the early champions of the Green Communities movement and to the collective work by so many to ensure that everyone has a home in a thriving community on a flourishing planet.

Contents

Acknowledgments

This book exists only because of the thousands of people who have dedicated their lives to ensuring that all of us have an affordable place to call home on a planet that is thriving. The community of people from diverse backgrounds and professions who are constantly pursuing better methods, materials, financing, and ways of working inspire me every day. I am ever so grateful to them; their voices and generous spirits compelled me to write this book so that more people know what is possible.

I was fortunate to have met Chuck Savitt, founder of Island Press, who found the story of Green Communities so compelling that he urged me consistently over the past 15 years to write a book. He introduced me to Heather Boyer, who patiently encouraged and assisted this process over the last many years. My colleague, mentor, and friend, Stockton Williams, has always been the one to think big and has had the will to make it so; I owe you a tremendous thank-you for your collaboration and endless co-conspiring.

Much like the integrative design process, the writing of this book was made possible because of the many people that have openly shared their diverse expertise and wisdom with me. The Green Team at Enterprise was pure magic. Thank you for your friendship, patience, and dedication, Ray Demers, Krista Eggers, David Epley, Yianice Hernandez, Amy Hook, Bomee Jung, Trisha Miller, Emily Mitchell, Josh Owens, Esther Toporovsky, and Diane Westcott, and honorary member Patty Rouse. A callout to every Enterprise local office Investment and Loan Fund team member, the Housing and Community Development Workgroup, Enterprise Homes staff, and the Orbiteurs, the Rose Fellowship, and Enterprise's Policy, Communications, and Events team, without whom we could not have galvanized a national movement. Each of you made the work not only possible but better.

I am grateful to everyone who took the time to talk with me over the years about the intersection of housing, health, and the environment. In particular those of you who spoke in depth with me as I was developing this book: Noreen Beatley, Michael Bodaken, Norma Bourland, Gina Ciganik, Madeline Fraser Cook, Paul Cummings, Chandika Devi, Ralph DiNola, Anne Evens, Jay Golden, Nicole Gudzowsky, Warren Hanson, Bart Harvey, Diane Hernandez, David Heslam, Joel ben Izzy, Michael Johnson, Gray Kelly, Joseph Kunkel, Jeff Lesk, Michael Levison, Deron Lovaas, Dr. Tiffany Manuel, Sunshine Mathon, Amanda McIntyre, Michelle Moore, Todd Nedwick, Ruth Ann Norton, Tom Phillips, Barbara Picower, Tim Robson, Nick Stenner, Wes Stuart, Katie Swenson, Charles E. Syndor III, Nick Tilsen, Ted Toon, Cheryl Wakeham, Bill Walsh, Peter Werwath, and Diane Yentel. Additionally, thank you to the many others with whom I have had the distinct pleasure of crossing paths over the years to advance housing and environmental justice for all.

Without a month at the Rockefeller Bellagio Center, I never would have had the time to lay the foundation for this book. Thank you to Pilar Palacia, the Rockefeller Foundation staff, and my fellow residents for creating such a generative space.

Carlos Lema, you deserve special mention. When we first met, I had the outline of this book taped up around my home office. Your support along the way enabled me to bring this project to life. Thank you.

Preface

By the time I was 17, I had moved 11 times. My family named each of the houses we lived in, typically after the street address—the Hancock house, the Cymbidium house, the Wokingham house, and so on. I have been back to many of those houses. When I visit them I stand on the opposite street corner, and memories, good and bad, flood back. Attachment to home is not a rich or poor sentiment. It is a human one. Everyone needs a safe and secure place to call home. In his 1944 State of the Union address, Franklin Roosevelt declared that the United States had a "Second Bill of Rights," including the right to a decent home. In 1948, the United States signed the Universal Declaration of Human Rights, recognizing adequate housing as a component of the human right to an adequate standard of living.[1] In the age of pandemics, climate change, and economic upheaval, a safe, secure place to call home takes on a whole new level of necessity.

We do not have the time or the resources to meet our housing crisis without also considering how to meet our climate crisis. I propose that we do both at the same time. Think of it this way: gray housing connotes practices that benefit a few in the short term but negatively impact the majority of us and our planet, whereas green housing connotes practices that benefit all of us and support the health of our planet now and for the long term.

Because of my own experience moving from place to place, I have spent the better part of the last several decades focused on ensuring that all of us have a place to call home. Like air to breathe and food to eat, housing is a basic human need. I have a deep understanding of the way a home can lift you up or how it can weigh you down. This is partly from the structure itself, how it was designed, constructed, and maintained, and partly from where it is located and whether it connects you to the community or isolates you. Our home is one of the building blocks of our life.

Throughout my career I have focused on many of the issues related to ensuring that everyone has access to a home that contributes to their overall health and well-being and to that of the planet. As a college student I volunteered with Habitat for Humanity, and almost 20 years later I helped construct and fund a net-zero, passive house for DC Habitat. I interned with the Oak Park Housing Center to track racist practices that were destabilizing neighborhoods, I witnessed the impacts of unhealthy and inadequate housing on girls' school performance while volunteering in the Peace Corps in Belize City, I examined regional planning barriers to better neighborhoods in Minnesota, I worked on behalf of seniors who wanted to stay in their community in a South Dakota farming county, I have engaged with whole communities to write housing action plans, and I have assessed statewide infrastructure needs and addressed land use and zoning issues that can strengthen communities. And for close to 10 years I shaped a national movement toward green affordable housing known as the Green Communities Initiative, while working for Enterprise Community Partners, a national housing and community development intermediary. This included developing the country's first and still only green building standard for affordable housing, the Green Communities Criteria.

The Situation

The United States is at a crossroads. We have at least 20 million people paying more than 30 percent of their income on housing, when "affordable housing" is defined as paying less than 30 percent of your income for housing. We are also approaching dangerously high levels of greenhouse gas emissions (since greenhouse gas emissions are often calculated as carbon dioxide equivalents, I will refer to them throughout the remainder of this book as "carbon emissions") in the atmosphere that, if not curbed, could lead to unalterable climate change. The US government cannot afford the costs of inaction as the health care costs mount from unhealthy housing and dirty energy production, as the population of unhoused persons grows, and as the impacts of climate change exacerbate federal and state budget deficits with each emergency. However, we can meet both our housing crisis and our climate crisis with green affordable housing. We have proof that green affordable housing is possible and for no additional cost. It can measurably reduce carbon emissions, improve human health, and support thriving communities.

I vividly recall being chastised in my final group project for an executive real estate program in which I had enrolled in 2007. We had been given an assignment to propose a redevelopment plan for several parcels of land at the water's edge in Providence, Rhode Island. Based on maps provided by Architecture 2030,[2] I had proposed a solution that called for the waterfront to be a park—an area that could be flooded without any material damage. Our professors and visiting experts criticized me for not maximizing the use of the most valuable waterfront property. They also told me that my proposition that the land was at risk for flooding would scare off any investors in the deal.

As the youngest, least-experienced developer in the room and one of only four women in the whole program, I was embarrassed. Years later I feel vindicated in my proposal. I know through experience that it would be negligent to build anything of permanence on those parcels that was not prepared to withstand substantial flooding. Almost 10 years since that day of judgment, I remain amazed that waterfront development continues in low-lying areas; in many communities, such residential neighborhoods are experiencing daily flooding from intense rain events.

Similarly, we have a vast body of evidence for the impact of polluted indoor air quality on human health.[3] I have become particularly attuned to the air quality of our indoor spaces since working to advance green affordable housing. Prior to doing this work I had no idea that our homes contained a barrage of toxic chemicals off-gassing and exposing us to known carcinogens, asthmagens, and endocrine disruptors. Since working for Enterprise Community Partners to launch and expand the Green Communities Initiative, I have changed many of my own practices, and I have lived in green housing. In 2012, I leased an apartment in a fairly new building on the Upper West Side of New York City that was certified as green by the United States Green Building Council. During that time I participated in a pilot study with the Environmental Defense Fund. I wore a bracelet that used wristband monitor technology from MyExposome, Inc., developed at Oregon State University, to detect hidden chemical exposures in my living environment. There were 27 other people in the study, and the technology could detect up to 1,400 chemicals. The wristbands were designed to "act like sponges to detect chemicals found in the air, water, and consumer goods." I was not too worried and agreed to wear the wristband.

The project detected 57 chemicals from the participant group. The average number of chemicals detected for any one participant was 15. Much to my horror my wristband detected 11 chemicals, including persistent, bioaccumulative, and toxic ones. They sound as scary as they are. Bioaccumulative chemicals persist for generations and accumulate in the body. One chemical was a polycyclic aromatic hydrocarbon, which is an air pollutant linked to cancer. I was exposing myself to hazardous phthalate and flame retardant chemicals, most of which are banned in the European Union due to their reproductive ill health effects. The wristbands of other people in the study detected the pesticide permethrin, which has increasingly been linked to neurological effects, including Parkinson's disease. Indeed, what you don't know can kill you.

Those life-threatening toxic chemicals do not show up inadvertently in building products; they are ingredients chosen to be in those products by the manufacturers. But manufacturers of building products and materials do not have to disclose these ingredients to the consumer. Out of the approximately 80,000 chemicals in products today only about 300 have been tested for health and safety, only nine of which have been banned by the Environmental Protection Agency (EPA). This is not because only nine are harmful but because the testing process is so ineffective.[4] Communities with manufacturing facilities where these building products or their ingredients are produced can be exposed to lethal levels of poisonous particulate matter every day.

We Can Stop the Crises

Max is an artist who lives in the beautiful Sawmill Community in Albuquerque, New Mexico. One day decades ago he woke up and rolled out of bed to start the day. He was startled to notice an outline of his body, in sawdust, on his bed. Pollution from the local particleboard manufacturer had infiltrated his home and found its way into his most personal of spaces. That experience and others like it prompted Max to become one of the early organizers who fought back against the pollution that was covering his community.

Connie Chavez, the former executive director of the Sawmill Community Land Trust, grew up in Albuquerque and told me the story of Max. She explained to me that the facility had unlined waste pits containing chemicals used in the manufacturing process. She and her friends often sneaked into the fenced area to play. These chemicals were not confined to the pits; they were also released into the air. Formaldehyde is a petrochemical derived from fossil fuel and is often used to bind the ingredients in particleboard and as a resin to make the board water resistant. Exposure to formaldehyde can trigger asthma, and it is known to be a carcinogen. The fossil fuel infrastructure developed in this country is contributing to a proliferation of petrochemicals like formaldehyde, many of which are showing up in our building materials.

The Sawmill story caught my attention when I first visited there almost 20 years ago as I was learning about green building. It seemed to be the embodiment of the Margaret Mead quote, "Never doubt that a small group of thoughtful committed citizens can change the world. Indeed it is the only thing that ever has." Once an apple orchard, the area became the location for lumber companies, sawmills, and a particleboard manufacturer. Because particleboard is made with waste materials like wood shavings and sawdust, it made sense that a manufacturer would co-locate there. Eventually the lumber companies left, but the particleboard company remained.

Many people in the community worked for the company, which made the decision about what to do with the facility particularly fraught with uncertainty. Ultimately the community coalesced to shut it down, and they won. The company ceased its operations and abruptly left town. The site was fenced off and sat as a reminder of the community's lack of value in the eyes of companies who could legally poison the air, soil, and water and then leave the site as a toxic waste dump, much like the coal and fossil fuel companies are doing today as they close their fracking wells, pipelines, mines, and power plants.[5]

Shutting down the facility was the beginning of a series of fights residents took to protect their community. As the community organized around the environmental issues, residents became increasingly concerned about encroaching development from downtown Albuquerque, which was threatening to push people out to make room for business and higher-end housing. The most active community organizers joined to form the Sawmill Advisory Council. To gain control of their community they formed the aforementioned Sawmill Community Land Trust and purchased 27 acres of land, including the particleboard manufacturing site, using funds from a $1.2 million Community Development Block

Grant made available by the State of New Mexico. This was a major victory for the community.

Today, the Sawmill Community Land Trust is one of the largest contiguous community land trust projects in the United States. Because of the history of the site and the community's commitment to social and environmental justice, it wanted to ensure that the development restored the health of the community. The Sawmill Lofts are live/work spaces intended for artisans and designed to meet the Green Communities Criteria. The Criteria prohibit the use of particleboard with urea formaldehyde because of what I learned from the Sawmill Community and the many communities that have been exposed to this toxic manufacturing process.

The story of Sawmill underscores the need to consider green development not only for the end product, the green house, but for the entire set of activities that allow it to be considered a green development. A house cannot be green at the expense of making another house or community gray. Because of the efforts of community organizers like Max, formaldehyde has been banned from use in particleboard, but we must still continue using our collective organizing capacity to end the cycle of pollution.

As I make edits to this manuscript, the world is fighting to stop COVID-19 from spreading. The source of the virus is not entirely known, but some suspect that as we cut down the forests needed to clean our air, curb pollution, and sequester carbon, we are more likely to come into contact with various species of bats and other mammals. These make their way into our food supply chain, and we become exposed to pathogens for which we have no immunity. The Center for Climate, Health, and the Global Environment at the Harvard T. H. Chan School of Public Health asserts the following: "Many of the root causes of climate change also increase the risk of pandemics. Deforestation, which occurs mostly for agricultural purposes, is the largest cause of habitat loss worldwide.

Loss of habitat forces animals to migrate and potentially contact other animals or people and share germs. Large livestock farms can also serve as a source for spillover of infections from animals to people. Less demand for animal meat and more sustainable animal husbandry could decrease emerging infectious disease risk and lower greenhouse gas emissions."[6]

I write this book as a thank-you to, and a reminder of, all the people who are working tirelessly to provide green housing in their communities and who fight for environmental and climate justice. I also write it as a call to action. Let us join together to transition from gray to green communities for everyone.

Introduction

THE OPPORTUNITY TO CHANGE THE COURSE OF human history is within our grasp. Over this decade we can finally deliver on the recognized human right to housing and reach a level of carbon emissions acceptable to scientists.

Here is what troubles me: At the very time we need to learn how to flatten the curve of a pandemic, take effective action toward racial justice, curb our carbon emissions, and prepare for the impacts of climate change, our nation is utterly underdelivering the housing we need. The only housing stock within reach for most people living in the United States is inefficient, expensive to operate and maintain, unprepared for intense weather events, and potentially unhealthy. This is unjust and unstainable, and it requires decisive action because housing is a human right and a basic necessity. This book is about how our country can do better so everyone can enjoy the benefits of safe, secure, and healthy housing.

The math is simple. We have a growing population, a shortage of housing, and incomes that are not keeping pace with the cost of living. This equates to working people being unable to access housing they can afford in communities where they need or want to live. It also adds up to a housing stock out of sync with our climate action goals. These older homes are demanding more energy than they would need if they were rehabilitated. Furthermore, new housing is not constructed in ways that significantly mitigate climate change. We are falling farther and farther behind in delivering the amount of housing needed at the price point affordable to most and are thereby weakening the foundation upon which society rests. When households must make untenable trade-offs between rent and food or medicine, between the mortgage and the utility bill, between our housing costs and education, we are undermining people's health, economic security, and the overall well-being of our country.

Every day that we postpone meeting our housing needs, we negatively affect other parts of society and allow people to fall farther behind in becoming housing secure. Our racialized land-use policies, financial mechanisms, and housing programs have perpetuated a disproportionate wealth gap in this country that is widening by the minute and bringing our society to the brink of collapse. We cannot function without ensuring that we have housing that all people can access and afford, and simultaneously addressing the existential threat of climate change.

A Green New Deal that prioritizes a green housing guarantee exemplifies one type of commitment to housing that we need. The Green New Deal started as an idea largely hatched by youth who had been involved in the fossil fuels divestment movement across college campuses.[1] Known as the Sunrise Movement, it proclaims to be a movement to stop climate change and create millions of good jobs in the process. It was based on an idea to address climate change at the scale that the crisis demands, not to benefit the elite but to benefit those on the frontlines. Members of the movement espouse to be

> building an army of young people to make climate change an urgent priority across America, end the corrupting influence of fossil fuel executives on our politics, and elect leaders who stand up for the health and well-being of all people. We are ordinary young people who are scared about what the climate crisis

means for the people and places we love. We are gathering in classrooms, living rooms, and worship halls across the country. Everyone has a role to play. Public opinion is already with us—if we unite by the millions we can turn this into political power and reclaim our democracy. We are not looking to the right or left. We look forward. Together, we will change this country and this world, sure as the sun rises each morning.[2]

If we fail to solve these crises now they will continue to worsen. They will not go away, and we are witnessing the results of not act-ing swiftly enough or robustly enough. On any given night in the United States, there are over 560,000 people of all ages experiencing homelessness. Our population of unhoused is growing because we are not providing housing at all levels of the income spectrum. In places where housing does not exist at an affordable price point, people whose employers are paying them low- and very-low incomes will find themselves allocating most of their paycheck for rent or a mortgage and utilities. Many have little of their income available for other necessities, or for unexpected expenses, such as those that occur during a pandemic. In fact, 8.1 million rental households report that they do not have the financial resources to evacuate their home and relocate if a disaster strikes. In 2018, 20.8 million of the total 44 million renter households were paying more than 30 percent of their income for rent and utilities,[3] and 10.9 million were paying more than 50 percent of their income for rent and utilities. That is, one in four of our 44 million rental households is spending more than half of its income on rent and utilities.

Renters seeking more affordable housing options might move in with other families by doubling or even tripling up, which has mental and physical consequences. Or people might move into homes that are located far from where they work, compounding the impact on that household's budget by increasing transportation costs and time away from family, in turn adding costs for child or elder care.

The barriers to homeownership are significant for low- and moderate-income households. As the Joint Center for Housing Studies at Harvard University points out in its *State of the Nation's Housing 2019* report,[4] we are not building enough housing to keep up with demand. As a consequence, many households are forced to live in older homes. According to the 2016 American Community Survey, the median age of all homes was 37 years, which means that 50 percent of all the existing homes are older than 40 years. These older homes are likely out of compliance with today's building codes and are quite likely in need of repair, structurally unsafe, and unhealthy, and they could burden the residents with higher utility and maintenance costs.

Housing as a Vaccine and Game Changer

Bernice Aquino rented a 1950s row home in the Belair-Edison neighborhood of Baltimore with her 12-year-old daughter, Jacquirelyn. The deteriorating rental property had chipping and peeling paint, pest issues, and other health and safety hazards. Jacquirelyn suffered from asthma, which often interrupted her sleep. This led to missed school days for Jacquirelyn and missed work days for Bernice. At the hospital, during one of Jacquirelyn's asthma attacks, Bernice was given material about making home improvements that could reduce asthma triggers. These materials were produced by the Green and Healthy Homes Initiative (GHHI).

GHHI is a Baltimore-based, nonprofit organization that advocates for healthy, safe, and energy-efficient homes. Bernice contacted them to schedule an initial home visit that included both an environmental assessment and an energy audit. The audit identified numerous hazards and weatherization deficiencies in the home, including lead hazards on interior and exterior surfaces; pest infestation due to gaps in the walls, floor, and under the kitchen sink; mold and condensation in the bathroom due to poor ventilation and a leaky bathtub faucet; air leakage around exterior doors, windows, rim joists, and plumbing leads; uninsulated pipes and water heater; lack of weather stripping; and an inefficient furnace.

Healthy home and lead hazard reduction interventions resulted in 10 lead-free, Energy-Star replacement windows; paint stabilization of doors, baseboards, and walls with integrated pest management; installation of bathroom exhaust fans and smoke and carbon monoxide alarms; mattress and pillow covers to reduce dust mites; HEPA vacuuming and wet cleaning of all floors to reduce indoor allergens; and distribution of a HEPA vacuum and indoor allergen cleaning kit. Weatherization measures included a furnace tune-up and cleaning, foam insulation to seal air leakages, an insulation jacket around the water heater tank, pipe insulation on the water lines, roofing repair to reduce water infiltration, weather stripping, and low-flow plumbing. GHHI conducted follow-up home visits for asthma case management and energy-usage education.

This has been a game changer for those in the healthy homes field who recognize the health benefits of making energy improvements, but until recently, they were only able to use available funding for direct health improvements. Doing so is a win-win-win—or as people in Baltimore might say, a "trifecta"[5]—for homeowners and families, for the local hospital that no longer has a repeat patient in its emergency room for a preventable disease, and for the energy utility, because the decreased demand allows it to serve more customers without expanding its power plant.

If these interventions had been done separately as lead hazard reductions, weatherization, and energy efficiency improvements, they would have cost an estimated $12,000. However, completed in one braided intervention approach, they cost less than $10,000. The savings resulted from the efficiencies gained by having one contractor manage all of the interventions and one entity manage the blended sources of funding to pay for the repairs. When this approach is replicated in over 100 homes, the savings add up quickly, allowing GHHI to complete many more whole-home improvements.

After the improvements were made (see Figure 0.1 for a photo of this Baltimore row house), Jacquirelyn was no longer waking up wheezing or coughing or depending on her rescue inhaler medication to get through the day. Jacquirelyn's health improvement after her home was renovated is not unique. Housing can be a powerful vaccine for health.

Gray to Green Housing

Bernice's home prior to the improvements can be described as "gray" housing. Whether it is old or new, gray housing serves only the immediate and necessary function of providing shelter. Gray housing does not deliver other benefits for residents, communities, and the planet. As already described, Bernice's gray home detracted from the health of her family and their capacity to access opportunity. Its inefficiencies contributed to climate change by requiring more electricity and gas to be generated from greenhouse gas–emitting fossil fuel power plants to heat and cool the home.

Figure 0.1. 3570 Dudley Avenue, Baltimore, MD (Photo credit: © Harry Connolly, 2020)

Like Bernice's home after the renovation, green housing addresses residents' needs and aspirations. It improves health; it conserves energy and water and therefore requires fewer carbon emissions; and it uses materials that are beneficial both for the residents and for the planet.

Diana Hernández and Carolyn P. Swope from the Columbia Mailman School of Public Health write in the *American Journal of Public Health*, "The links between housing and health are now known to be strong and multifaceted and to generally span across four key pillars: stability, affordability, quality and safety, and neighborhood opportunity. Housing disparities in the United States are tenaciously patterned along axes of social inequality and contribute to the burden related to persistently adverse health outcomes in affected groups. Appreciating the multidimensional relationship between housing and health is critical in moving the housing and health agenda forward to inspire greater equity."[6]

The promise of green housing can extend to the health of those who live or work along the housing material supply chain. Producing building materials without using toxic classes of chemicals, many of which are derived from fossil fuels, will improve the health of the workers and the environment of communities where those materials are manufactured and through which they are transported. That is, all of us working across the real estate sector—from policy, to finance, to construction, and to operations and maintenance—must think about the health impacts from the extraction of ingredients and the production of materials, not just their end use in the construction and finishes of the homes. In this way, we are extending the benefits of green building from the people living and working in those properties to millions of others involved in the manufacturing of materials or living in communities with the manufacturing facilities. We are also reducing the carbon emissions involved in extracting, manufacturing, and transporting those materials.

If the extraction, production, and transportation of materials and products had been healthy from the start, then we would be closer to having truly green housing. We could realize a safe and "circular economy," in which we design out waste and pollution and rebuild natural capital and keep products, materials, and molecules flowing safely and effectively through the economy at their highest value.[7] Some materials should not be used because they contain substances that negatively affect human health and the environment. In other cases, the way in which materials are combined in a product inhibits their separation and capture after use, resulting in additional waste and lost economic value. To achieve a circular economy, in which products cannot accumulate in a landfill or aggregate in the ocean, our goal must be to create both a healthy building material supply chain and a healthy product and home.

When we engage in the powerful synergistic combination of (1) the energy efficiency delivered to the resident in the green home and (2) the health of the community where the energy efficiency products were sourced, we create a powerful collaboration among advocates for the green economy and environmental justice groups. For example, in 2005, factories manufacturing fiberglass insulation in the United States and Canada released nearly 600,000 pounds of formaldehyde into the air and nearby communities. As a result of market pressure from the environmental justice and public health advocates for healthier alternatives, residential fiberglass insulation manufacturers in the United States and Canada replaced formaldehyde-based binders with less hazardous alternatives, leading to a 90 percent drop in formaldehyde emissions from these facilities as of 2011.

Embrace the Complexity

The climate and housing crises are complex. There are no singular answers. We need big, mutually inclusive, moonshot actions on both fronts. We must have a national commitment to ending our use of fossil fuels, and we must have a national commitment to ensuring that everyone has a home they can afford. As we address both crises we must start by asking who will benefit. Who is part of creating solutions and who is not? Who decides how we proceed? Let us not squander this opportunity to make massive investments in ways that repair society and advance us toward a just and equitable future.

For far too long, housing has been the physical demonstration of the racist systems embedded in our country. Inherently our policies, programs, and practices that extract wealth, knowledge, and life for the profit and gain of a few have abandoned whole races, peoples, and communities. That same thinking has brought us to the brink of destroying our planet. The thesis of this book is that we are connected to one another and to our planet. We must address this country's racial disparities to fully realize the benefits of a green way of thinking and acting. If we do not, we will perpetuate systems that have marginalized communities of color. We must recognize the inherent injustice embedded in our current housing system that stems from colonialism and White supremacy.

Land sovereignty recognizes the right of the community to have effective access to, control over, and use of land. It remains one of the least realized, and least discussed, aspects of green housing. There are a few efforts working to change that, such as those of the Sustainable Native Communities Design Lab at MASS Design Group, which grew from a collective belief that each community must determine its own path toward sustainability; NDN, a nonprofit whose mission is to build the collective power of Indigenous peoples, communities, and nations to exercise their inherent right to self-determination while fostering a world that is built on a foundation of justice and equity for all people and the planet; and the Center for Heirs' Property, which promotes the sustainable use of heirs' property when property owners—predominantly Blacks—die without a will, often resulting in a family's loss of the land. If we are to have truly green communities, the public sector and community development professionals must stop making decisions for people about the development of their communities and the locations where they should live. The rest of us must do the same.

Although the price of land may be one of the most prohibiting factors in meeting our housing demand, the true cost of this land transcends our current transactions. We must recognize the history of the land, and we must listen to the needs and aspirations of communities, past and present, to enable the type of housing and development they desire to have. We must realize that without access to land and housing ownership, Indigenous people, Blacks, people of color, and immigrants have been denied the ability to accumulate wealth and pass that wealth on from one generation to the next. "For every crane that's going in the air, you should ask, 'Is this building a pathway out of poverty, or trapping somebody into debt?'" asks Odessa Kelly from Stand Up Nashville.

Our Nation's Future Is at Stake

In 1992, at the Rio de Janeiro Earth Summit, a 12-year-old delivered remarks that would gain her notoriety as, "The Girl Who Silenced the World for 5 Minutes."[8] Severn Cullis-Suzuki used her time on the global stage to speak on behalf of all generations to come, to reprimand the adults present. She was not referencing climate change specifically but the broader environmental concerns that come from pushing the limits of the Industrial Revolution and a consumer-based society. She declared to the global leaders that "if they don't know how to fix it, [then] stop breaking it," referring to the health of our environment. She said she was afraid to breathe the air because she did not know what was in it and that she was afraid to go out in the sun because of the hole in the ozone.

Greta Thunberg, a 16-year-old climate activist from Sweden, started a global, student-led climate strike movement. In her 2019 remarks at the UN Climate Action Summit, she reprimanded us with these words: "This is all wrong. I shouldn't be standing here. I should be back in school on the other side of the ocean. Yet you all come to us young people for hope? How dare you! You have taken away my dreams and my childhood with your empty words. And yet I'm one of the lucky ones. People are suffering. People are dying. Entire ecosystems are collapsing. We are in the beginning of a mass extinction. And all you can talk about is money and fairy tales of eternal economic growth. How dare you!"[9]

Children are powerful change agents. I agree with Severn and Greta that the burden should not fall to them. Even if we do not have all the solutions to sustain our planet, we must take action on what we do know while we address the root causes and cocreate solutions. We must, as Severn demanded, "stop breaking it."

Gray housing is breaking our communities and the environment. It is unsustainable and unconscionable to continue as if we do not know better. We must transition from gray to green in ways that deliver health, economic, and environmental benefits for everyone, starting with those who stand to gain the most and who had the least to do with creating the crises in the first place.

CHAPTER 1

The Problem with Gray

How do we deal with abandonment, ruin, decay? How do we start to imagine ourselves as deeper caretakers of the things that exist in the world?

—Theaster Gates

OUR HOUSING SYSTEM IS BROKEN. IT IS underperforming and in need of disruption. It is a fragmented array of sectors, policies, and programs that are undermining the success of our society by leaving millions without housing they can afford and hundreds of thousands without a home at all. In our current housing system, mortgage, rent, and utility costs are out of reach for most persons in America. Housing that is attainable is often older and in outlying areas that are not easily accessible to jobs, quality schools, and services. Gray, or conventional, housing is out of line with our mandate to curb carbon emissions. As already described, it diminishes human and planetary health. The gray housing system has been used to create a massive wealth gap in this country and to segment society into the haves and the have-nots and by racialized notions of who is creditworthy and of value.[1] I am not proposing that we simply supply more housing in a hurry as we have

done in the past to spur the economy or house some of our return-
ing war veterans. That mind-set contributed to where we are now,
with expanding income inequality, an explosion of people who are
unhoused, and escalating carbon emissions.

Gray housing detracts from our ability to participate more fully in
our communities. Housing affordable to all is recognized by the United
Nations as a basic human right.[2] We are failing to realize it. Not doing so
is having ripple effects across the country because housing is inextricably
linked to every other aspect of our society.

I remember a story Geoffrey Canada, president of the Harlem Children's Zone, told during the Picower Institute at MIT's 2018 spring
symposium on toxic stress. He talked about a young boy who attended
school within the Harlem Children's Zone and participated in an
after-school program. He was reportedly hearing voices and acting
strange. The teachers wanted to take him to the doctor, but Geoffrey
knew that could have consequences, so instead a teacher went to the boy's
home to better understand the situation.

The teacher found that the apartment was infested with rats. The
mother said it was the only place she could afford and that if she said
anything to the landlord, she was afraid she might lose the apartment. It
was the boy's job to stay up at night and make sure the rats did not bite
his sisters so the mother could sleep and go to work. The boy had not
had a decent night's sleep in weeks. If he had been sent to the doctor, he
probably would have been put on medication for mental illness, but all
the family needed was a home where they could live securely, with dignity
and without worry. This one point in the boy's life could have sent him
down a very different path than the one ultimately provided by improved
housing. Gray housing is needlessly harming children's lives and taking a
toll on the ones who love and care for them.

The following sections address the impacts of gray housing on people
and the planet, including high housing costs, unaffordable utility bills,

undesirable locations and housing, and carbon emissions. Furthermore, the gray housing sector is itself fragmented.

High Housing Costs

Mortgage and rents are out of reach. The number of people unable to purchase or rent a house they can afford is growing; new housing is not built at the rate necessary to keep up with household growth, which limits supply and increases prices. The widely accepted definition of "affordable" is when a person or household pays less than 30 percent of their income for housing and utilities. The United States has approximately 138 million housing units available, and about one-third, 44 million, are renter occupied.[3] In 2018, 20.8 million renters were paying more than 30 percent of their income for rent and utilities, with over 11 million paying more than 50 percent of their income. Of the 11 million about 7.5 million have extremely low incomes, meaning they earn incomes at or below the poverty line or 30 percent of the area median income, whichever is greater, and we are not providing enough rental assistance. Only 4.75 million of the 19 million households eligible for rental assistance receive it. If trends continue, middle- and lower-income households will be caught in the untenable positions of paying the majority of their paycheck for rent or a mortgage and will not be able to afford other basic needs, including food, water, electricity, internet and cell phone service, and transportation as shown in Figure 1.1. Families not being able to afford housing is a problem not only for them but for all of us because it destabilizes communities. We see this in communities where teachers cannot afford housing because their wages are too low compared with the cost of housing, making it extremely difficult for the school district to attract and retain a workforce and in some cases leading to teacher strikes.

In 2019, 567,715 people in the United States lacked a fixed, regular, and adequate nighttime residence.[4] Additionally, 4 out of every

Monthly Expenditure (Dollars)

Expenditure Categories

Figure 1.1. Monthly expenditures (Source: Joint Center for Housing Studies Tabulations of Bureau of Labor Statistics, 2017 Consumer Expenditure Survey)

10 people experiencing homelessness were Black or African American (40 percent or 225,735 people) despite being only 13 percent of the population. We are becoming resigned to the high cost of housing and the growing numbers of people who cannot afford to pay, many of whom are working. It is possible to provide access to housing that is affordable for all in places where it is needed. A few cities have committed to doing so, and these models can be replicated. Virginia, for example, was the first state to reach a "functional" end to homelessness among veterans.[5] It started with research and a challenge to end veteran homelessness in four communities, which expanded into a successful statewide effort.

"There's a cruelty here that I don't think I've seen," said Leilani Farha, then the United Nations special rapporteur on adequate housing, after a 2018 visit to Northern California. She compared conditions there to those in countries that, unlike the United States, lack the money to care for their citizens.[6] How unaffordable is housing in the United States? A full-time worker earning the federal minimum wage of $7.25 needs to work 99 hours per week for all 52 weeks of the year, or approximately two and a half full-time jobs, to afford a one-bedroom home at the national average "fair" market rent—and if you think the only people making minimum wage in this country are teenagers and retired persons, think again. In 2017, according to the Bureau of Labor and Statistics, 80 million workers aged 16 and older in the United States were paid at hourly rates, representing 58 percent of all wage and salary workers. Among those paid by the hour, 542,000 workers earned exactly the prevailing federal minimum wage of $7.25 per hour. We now know these workers as our "essential" community members—the ones selling groceries, cleaning hospitals, driving buses, teaching our children, and delivering packages bought online.

To make housing costs more affordable, families often double up and crowd into places designed for one or two persons not five or six. This places a burden on the structure, which has to provide more electricity, water, and wastewater services than it was designed to support. It also strains communities' capacity to provide infrastructure and services because the increased number of people were not part of the planning and zoning process. Higher density, on the other hand, is when communities plan for more compact development and can provide adequate amenities and services to meet the needs of the people who will be living there. Overcrowding may also put households in violation of their lease, which could lead to eviction, increasing the future challenge of finding secure housing. Matt Desmond makes this point vividly clear in his book *Evicted*, in which he shows how people without secure affordable housing live on a ledge between homelessness and joblessness as they balance the need for shelter with the need to get to work and take care of their families.

In their 2019 State of the Nation's Housing report, the Joint Center for Housing Studies states the following:

If current housing supply trends persist, house prices and rents will continue to rise at a healthy clip, further limiting the housing options for many. To ensure that the market can produce homes that meet the diverse needs of the growing US population, the public, private, and nonprofit sectors must address constraints on the development process. And for the millions of families and individuals that struggle to find housing that fits their budgets, much greater public efforts will be necessary to close the gap between what they can afford and the cost of producing decent housing. The government calculates that $600 is the most a family living at the poverty line can afford to pay in monthly rent while still having enough money for food, health care, and other needs. From 1990 to 2017, the number of housing units available below that price shrank by four million.[7]

Unaffordable Utility Bills

For most people in this country, water is like air in that we think it is abundant and free. The fixtures, toilets, and appliances in gray housing are generally installed to meet required standards with little consideration for where the water is coming from, how clean it is, and what it might cost. On average, in 2018, a family of four paid about $70 a month for water. The price index of water and sewage maintenance has increased in recent years as infrastructure continues to age across the United States and local governments struggle to maintain it.[8] The owner or renter acquires the responsibility for paying the costs associated with supplying water to their house and land when building a property or deciding to move into one.

The other housing-related cost not typically bundled into a rent payment and not part of a mortgage is the cost of electricity or gas. The percentage of gross household income spent on home-related energy bills is considered the occupant's energy burden. According to the US Census

(2011–2016), the national average energy burden for low-income households is three times higher than for non-low-income households (see Figures 1.2a and b). Of all US households, 44 percent, or about 50 million, are defined as low income. The burden is not a result of solely having less income but is often the result of households living in older homes with inadequate insulation, inefficient equipment, and older appliances. When these types of homes receive new appliances, greater levels of insulation, and upgraded heating and cooling systems, they can save upward of $300 a year in utility bills. In the spring and summer of 2020 households impacted by COVID-19 may already have been struggling to pay energy bills, which became even more difficult to afford as people lost jobs and were told to shelter at home, which in turn led to increased use of electricity. In California, residential electric usage during the COVID-19 pandemic increased 15–20 percent compared to the same period in 2019.[9]

Families with high energy burdens may limit energy use to keep costs low by inadequately heating, cooling, or lighting their homes, resulting in unhealthy living conditions. Crystal Barbour from Charlottesville, Virginia, rents an apartment for herself and two children. She dreaded the monthly ritual of opening her utility bill before her home was made more energy efficient. "You would just eat cheap when it was electric bill week."[10] After energy efficiency improvements were made to her home—including efficient water fixtures, attic insulation, and lighting—her bills became affordable all year round. Programs like the Low Income Home Energy Assistance Program (LIHEAP), which can provide families like Crystal's with bill assistance, are underfunded and leave millions of energy-burdened households without this critical safety net. This results in deaths when people are trying to save money on their utility bill or when the utility cuts off power or water due to nonpayment. The report *Lights Out in the Cold* details this problem and includes many harrowing stories of families facing loss of power and water due to an inability to cover the costs.[11] Instead of bill assistance we must redirect

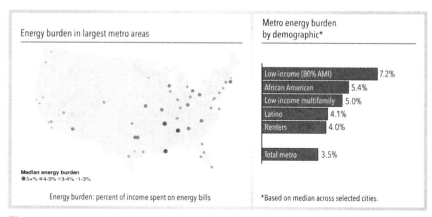

Figure 1.2a. America's energy burden (Source: A. Drehobl, L. Ross, and R. Ayala, *How High Are America's Energy Burdens? A National and Metropolitan Energy Burden Assessment* [Washington, DC: American Council for an Energy-Efficient Economy, forthcoming])

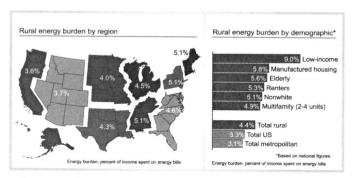

Figure 1.2b. America's rural energy burden (Source: L. Ross, A. Drehobl, and B. Stickles, *The High Cost of Energy in Rural America: Household Energy Burdens and Opportunities for Energy Efficiency* [Washington, DC: American Council for an Energy-Efficient Economy, 2018], www.aceee.org/research-report/u1806)

utility program dollars and other programs to pay for energy efficiency improvements so that we see permanent reductions in utility bills. This is one of the goals of the Energy Efficiency for All initiative and its social impact network, NEWHAB, that seeks to build collective power to ensure that all renters live in homes that are affordable and healthy.[12]

Undesirable Locations and Housing

Housing that is attainable is often in an unhealthy location. People who are not earning wages that keep up with fair market rent prices often end up in older units where rent is less than it would be in fair market communities that are walkable and healthier, with basic amenities like quality parks, grocery stores with fresh food, places of worship, cultural centers, and well-performing schools. In the United States there are vast differences of life expectancy between zip codes, due mainly to chronic diseases as well as homicides and injuries. Researchers from the New York University School of Medicine found that "56 of the US' 500 largest cities are home to people who can expect to live at least 20 fewer years than those in other neighborhoods, even if they're just blocks or miles away."[13] The researchers found that the cities with the biggest differences were also those with the most segregation by race and class. Where we live matters, and we live where we do because of racialized policies and practices that have generationally favored the White and the already privileged. Zip codes of the privileged have better housing conditions, less pollution, more green space, and greater opportunity.

Perhaps even more damaging can be the long-term effects on children growing up in zip codes with less opportunity and without safe and adequate housing, which expose children and their mothers to adverse childhood experiences (ACEs). ACEs often result when households are overcrowded or when caregivers have long or unpredictable commutes. Harvard's Center on the Developing Child and Dr. Nadine Burke Harris's book and TEDMED talk have awakened pediatricians and parents to the impact of ACEs on children and the need to pay as much attention to children's physical environments as to their nutrition and education.[14]

Children exposed to multiple ACEs may ultimately engage in risky health behaviors, face chronic health conditions, achieve a lower life potential, and even die earlier. The more ACEs a child has, the greater the chances of long-term physical and behavioral health issues. Six or more ACEs drop a child's life expectancy by 20 years. Just one ACE increases a child's risk for asthma by 28 percent, and four ACEs increase the risk by 73 percent. As Dr. Mona Hanna-Attisha writes in *What the Eyes Don't See*, "This new understanding of the health consequences of adverse experiences has changed how we practice medicine by broadening our field of vision forcing us to see a child's total environment as medicine. We are looking for larger factors in the child's world that can impede development and diminish an entire life and may put her at risk as an adult for diabetes, heart disease, or substance abuse. This is the most important concept in pediatrics and public health today."[15]

Living in Pollution
Among the underlying conditions putting people's health at risk is pollution. Harvard University released scientifically significant research in April 2020 providing evidence that "a small increase in long-term exposure to particulate matter leads to a large increase in the COVID-19 death rate."[16] People living in communities with power plants, peaker plants (power plants that run only during peak demand for energy, which in some communities means they run and emit high levels of pollution all summer), incinerators, diesel truck traffic, highways, and landfills can be exposed to dangerous levels of particulate matter.[17] COVID-19, a virus that attacks the respiratory and cardiovascular systems, will be more dangerous for those breathing polluted air on a daily basis.

Coal plants are single-handedly responsible for a large proportion of toxic emissions that poison local communities in the United

States. In a 2009 report to the Supreme Court, the Environmental Protection Agency (EPA) reported that coal power plants tend to be disproportionately located in low-income communities and communities of color. Tribal communities are also often burdened with nearby coal plants. In 2009, people who lived within three miles of a coal power plant had an average per capita income of $18,400, which was lower than the US average per capita income at that time of $21,587. Among those living within three miles of a coal power plant, 39 percent were people of color—a figure that is higher than the 36 percent proportion of people of color in the total US population.[18]

The environmental justice movement was partly launched in the early 1980s over a toxic landfill site in Warren County, North Carolina.[19] In 1983, a US Government Accountability Office report confirmed that there was a direct correlation between race and toxic waste sites. These connections are laid out in the 2016 report, *Coal Blooded: Putting Profits over People.*[20]

In Houston, you can take a "toxic tour" of the 30 or more refineries and chemical plants that have been allowed to locate next to homes and schools. From the east side of Houston to the Gulf of Mexico, you can witness the largest concentration of petroleum refineries and petrochemical companies on the planet. Juan Parras has been organizing there for decades and helped establish the Texas Environmental Justice Advocacy Services (TEJAS). These communities have been lied to over and over again. They are told the pollution is not harmful, they are told to stay inside their homes, and they are intimidated by industry when they raise their voices demanding justice. So why do they not just move? "Why should they?" is my answer. This is their home. Make the industry stop polluting. No community deserves to be sacrificed so that a few can profit at the expense of our collective health.

As Dr. Robert Bullard has said, even though Houston has no zoning, communities of color have unofficially been zoned as a compatible use with pollution. When Hurricane Harvey hit Houston in 2017, the conditions of air pollution were compounded with toxic flooding. Researchers at the University of Michigan and the University of Montana found "a consistent pattern over a 30-year period of placing hazardous waste facilities in neighborhoods where poor people and people of color live." They concluded that racial discrimination in zoning and the housing market, along with siting decisions based on following the path of least resistance, may best explain present-day inequities.[21]

Landfills

Landfills present potential threats to both the environment and human health. Although landfills are lined to protect the surrounding environment, malfunctions can still occur, and leakage is common. Those working or living around a landfill with leakage face hazards such as an increased risk of cancer and birth defects due to perilous airborne chemicals in both active and inactive landfills. As noted earlier people with low incomes and communities of color are more likely to live near a landfill and to have a landfill located by a private or public agency in their neighborhood. According to the Construction & Demolition Recycling Association, construction and demolition waste, which includes concrete, asphalt, wood, drywall, metals, asphalt shingles, and many other materials generated during building projects, is created at a rate of nearly 480 million tons per year, making it the largest individual waste stream in the country. The volume of construction waste generated worldwide every year, according to a report from Transparency Market Research, will nearly double to 2.2 billion tons by the year 2025. In the United States alone, an estimated 251 million tons of consumer solid waste is generated annually, but less than a third is recycled or composted. Furthermore, as much as 40 percent of this waste comes from construction projects, which produce a surplus of unused building materials.[22]

In Minnesota, construction debris is contaminating groundwater. Because construction waste in landfills is not included in the state's solid waste rules, the Minnesota Pollution Control Agency is pushing for tougher standards for demolition landfills that provide no barrier between deposited materials and groundwater. However, county officials across the state are pushing back against proposals to tighten regulations until the agency can pinpoint what exactly is contaminating groundwater.

Isolated Locations

Gray housing does not prioritize smart locations and is often located far from where people work and obtain services. Building housing in communities where land is often cheaper may further disadvantage low-income households because they may not have access to family-sustaining job opportunities, efficient and reliable transportation options, or adequate schools and services.

The transportation sector accounts for the largest share of carbon emissions,[23] which has escalated over the past decade as more households are forced to "drive till they qualify." A study by the Center for Housing Policy found that families earning $20,000–$30,000 spend nearly half their income on housing and transportation costs combined because they must drive away from job centers to where they qualify for housing that they can afford.[24] The housing and transportation affordability index is an innovative tool that measures the true affordability of housing based on its location. For residents living in location-inefficient neighborhoods it requires more time and money, while emitting more carbon emissions, to meet everyday mobility needs. These costs from location inefficiency can add up for households and their communities. Transportation costs can range from 8 percent of household income in location-efficient neighborhoods to over 26 percent in inefficient locations.[25] A home, no matter how affordable to rent or own, in a nonefficient location can exacerbate rather than ameliorate the financial struggle for persons paid a low income while also worsening pollution.

You might remember the story of 56-year-old James Robertson that became an internet sensation in 2015. Robertson's extraordinary story caught the heartstrings of most Americans. He worked in a factory over 35 miles away from his home in Detroit, earning $10.55 an hour. The buses did not get him all the way to work so he walked a total of 21 miles each day, a 12-hour round trip, never missing a day and never late. This sounds extreme, and it is, but it is not out of the ordinary for millions of people. Not locating affordable housing near reliable public transportation or job centers means that many people's lives and incomes are consumed by getting to and from work, hours of unpaid time. Additionally, fossil-fuel-burning transportation methods are a significant contributor to climate-changing carbon emissions.

Living in Harm's Way
Generally speaking in this Anthropocene epoch (the current geological age, viewed as the period during which human activity has been the dominant influence on the climate and the environment), there are few controls on what must be done on a site to account for the impact of constructing a house. This includes allowing landscaping that is incongruent with the weather and climate conditions of the region. Local controls may dictate what needs to be done on-site to address local conditions, but generally public agencies do not want to stand in the way of development. This is particularly true when it comes to developing in floodplains. It is generally accepted to do so. Development in 100-year floodplains is occurring at a slightly faster rate than in non-floodplains.

Living Far from Nature
The Centers for Disease Control and Prevention recommends widening sidewalks, connecting systems of paths for pedestrians and cyclists, and

creating community gardens to combat obesity and other health issues. The Trust for Public Land and its partners the Urban Land Institute and the National Recreation and Parks Association also call for everyone to have a park within a 10-minute walk of where they live because of the multiple benefits that parks provide, particularly to our physical and mental health. However, when sidewalks are crumbling, streetlights are not functioning, and air quality is unhealthy to breathe, it is unlikely that people will be out walking and accessing what little nature might be found in their communities.

Gray housing is often just good enough for now. A system that allows this to continue is broken because where you and I live can dictate the quality and length of our life. Too often our nation thinks it is enough to provide shelter for people who are trapped in our economic system, which is structured to benefit only a few. It would seem, and I have heard people say, that our communities cannot provide affordable housing that is "too nice." There is a pervasive myth in America that if people work harder or go to college, then they will earn more and can move into homes that do not require subsidies to keep the rents affordable. A lack of housing affordable to all is not the aggregation of personal failure. People live in housing that they can afford, and sometimes that housing is not healthy or not smartly located. We forget that an ill-designed home can hinder a person from accessing the opportunity that will improve their economic situation. As described earlier, our living environments can actually make us sick or can result in living arrangements that prevent us from following cultural customs important to our well-being.

Toxic Materials

The COVID-19 pandemic has shown us that structural racism is indeed a preexisting health condition. Racism structures opportunity and assigns value such that it plays out physically in our communities. Through our racialized land use and housing policies, we have allowed communities with majority populations of persons of color to become our dumping grounds, exposing people to toxic chemicals in the air they breathe, the water they drink and bathe in, and the soil in which they grow food and where their children play.

Nearly 6 million households live with moderate to severe housing hazards, including lead paint, water intrusion, injury and safety risks, pests, and electrical deficiencies. Moreover, 40 percent of asthma episodes are caused by preventable triggers in the home. This represents $5 billion lost annually in preventable medical costs. Over 5.5 million children and 19 million adults suffer from asthma in this country.[26] Persons with asthma are more vulnerable to other illnesses, such as COVID-19. In New York City, there is a huge disparity in asthma rates between more affluent and low-income communities, specifically in the Upper East Side of Manhattan, where asthma rates are 7 percent, and in East Harlem, where asthma rates are as high as 19 percent. This is an outrage. These children live only blocks away from one another and breathe the same outdoor air; it seems clear that indoor living environments are a major factor.

Beyond the usual suspects of moisture, mold, and pests, the indoor air quality of gray housing affects our health because of the materials used in construction and finish work. These toxic chemicals are rarely broadcast as being harmful, let alone easily identifiable on the labels of many products. They are allowed to harm our health without our permission or even our knowledge.

An architecture and design firm determines the configurations of someone's living space and creates a materials list for the construction team to follow as they build to the architect's plans and specifications. Those materials may contain toxic classes of chemicals that could trigger asthma, disrupt a person's endocrine system, be carcinogenic, or be neurogenic and could curtail a small child or baby's brain development. We generally assume that our federal government is regulating these products, but that is not the case. The Toxic Substances Control Act of 1976 grandfathered in 62,000 chemicals without any evaluation of their harm to human health. The Environmental Protection Agency assumes chemicals are safe unless proven otherwise and has a high threshold for proving harm. This is the converse of how chemicals are viewed in the European Union, which takes a precautionary stance and places the burden on the companies to prove that the chemicals are safe.[27]

Numerous building materials and products contain toxic classes of chemicals, such as highly fluorinated chemicals, antimicrobials, flame retardants, bisphenols, and phthalates that accumulate in our bodies. Little is known about the aggregated impact of these exposures on human health. In a tight living environment where children are exposed to these products, it is highly probable that they may encounter learning difficulties, either because the chemicals are neurotoxins or because they miss excessive days of school from the related illnesses these chemicals trigger, such as asthma, rashes, headaches, and stomachaches. Furthermore, as described in the introduction, the health effects from the exposure of these chemicals are not limited to people in their homes but extend to the workers manufacturing them and the communities where those manufacturing facilities are located, including air, water, and soil exposure to harmful by-products.

A study designed to assess materials used in a gray retrofit noted, "These materials often contain persistent, bioaccumulative, or toxic chemicals and either show evidence or are suspected of being asthmagens, reproductive or developmental toxicants, endocrine disruptors, or carcinogens. Not only are a building's residents endangered, but these chemicals of concern can also pose threats over the materials' life cycles to the workers who manufacture, install, and dispose of these products, to the communities adjacent to these facilities, and to the broader environment."[28]

A team of scientists looking at flame retardant chemicals in insulation noted, "Both air and moisture move through a building fabric, regardless of how tightly they are constructed. Substances within building cavities have the potential to migrate out of those cavities via movement driven by air, liquid and/or water vapor that occurs due to temperature, air and vapor pressure differentials. Chemicals may be present in dust from abraded materials or could volatilize and then settle in indoor dust to which building occupants could be exposed."[29]

In the COVID-19 pandemic people had no choice but to shelter in their homes. For some this meant increased exposure to harmful chemicals present in their homes, triggering asthma and making them more susceptible to respiratory illnesses like COVID-19. It was no surprise that disparities in death rates began to fall along racial and economic lines.

Across the United States homes with insulation containing formaldehyde, a known carcinogen and respiratory irritant, had significantly higher indoor levels of airborne formaldehyde that were associated with adverse health impacts for occupants. In the state of Washington, a worker in his mid-30s developed occupational asthma from installing spray foam insulation in residential attics and was forced to leave his job, a not uncommon occurrence. Spray foam insulation contains isocyanates, chemicals that can cause asthma and are toxic to the respiratory system. Isocyanates are a leading cause of occupational asthma, resulting in high economic costs to society.

African American children are five times more likely to be poisoned by lead because of our collective disinvestment in communities with majority Black populations. We knowingly allow children and expectant mothers to live in public housing that contains lead in the paint, in the water, and in soil producing the food that babies and adults alike eat. Lead is a poisonous and powerful neurotoxin that disrupts brain development. As the poster reads in my office, "There Is No Safe Level of Lead." According to the Centers for Disease Control and Prevention, roughly 500,000 children in the United States have elevated blood levels of lead. A child's main risk of lead poisoning comes from the lead-based house paints in near-universal use before 1950. The paints were banned for housing use in 1978. An estimated 24 million houses and apartments—which some 4 million young children call home—still have deteriorated lead paint contributing to lead-contaminated house dust. Unlike exposure to other toxic products, exposure to lead does not result in immediately identifiable symptoms; it accumulates, causing permanent damage. Although we think of lead mostly as permanently harming babies and young children, it also impairs the health of adults.

As we move on to stop newer toxic classes of chemicals from harming us, we must remember that millions continue to be poisoned by lead. As Jim Rouse, a developer and cofounder of Enterprise Community Partners, was known to say, "Whatever ought to be, can be." Lead poisoning ought to and can end if we have the will to make it so. Lead is nondiscriminatory in whom it harms, but our housing system has predominantly directed low-income persons, Indigenous people, and people of color into deteriorated housing, where lead paint still exists.

There are efforts underway to eradicate lead poisoning, but they are underresourced and lack the scale of capital necessary to tackle the problem. It is maddening that we have not committed to ending childhood lead poisoning. The longer we wait to do so, the clearer it becomes that our policies and programs do not value the lives of the children, predominantly those of color and those living in low-income households, who continue to be exposed. We must have antiracist housing programs that eradicate the unequal conditions leading to shorter life spans and health disparities for Indigenous people, Blacks, and all people of color.

Carbon Emissions

Buildings are responsible for 40 percent of the world's carbon dioxide emissions, 25 percent of the world's timber harvest, 16 percent of the freshwater withdrawal, and 30 percent of the raw materials produced in the United States. Residential and commercial buildings are responsible for approximately 29 percent of US carbon emissions. Without an explicit intention to consider these impacts, we will be unable to meet our goals to significantly curb carbon emissions, and we will continue to have undesirable outcomes related to human health.

There are many ways our housing contributes to climate change. Primarily it is caused by fossil fuel combustion to produce the energy we use for heating and cooling spaces, warming our water, and firing up appliances. From 1990 to 2015, carbon emissions attributable to the residential sector increased by 20 percent.[30] Additionally, the manufacture of products we use in our homes requires energy and natural resources as well, significantly contributing to overall carbon emissions. According to the United Nations Environment Programme, the global manufacturing of building materials accounts for 11 percent of our total carbon emissions. We are in a race to reduce emissions before we lose the opportunity to do so. The Intergovernmental Panel on Climate Change urges us to reduce total carbon emissions before 2030 in order to remain under the 1.5 degree Celsius threshold of additional warming. If we do not, climate change will be irreversible, and the rising costs to rebuild after disasters will far exceed the costs of taking action now (Figure 1.3). Although we can continue tweaking our housing sector to make sure it is not contributing to carbon emissions through our energy consumption, if we ignore the embodied carbon of building materials (the amount of carbon emissions from extracting, manufacturing, and transporting them for new construction or rehabilitation purposes), then we are locking in carbon emissions that cannot be reversed.

When it comes to climate change, we know Mother Nature does not discriminate, but humans do. Our discriminatory housing, land use, and labor practices and systems have put people of color and people earning low wages in vulnerable situations. We have seen this play out as the change in our climate accelerates and affects our weather, with the

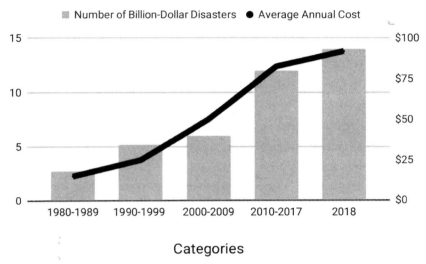

Number of Billion-Dollar Disasters and Average Annual Cost

Figure 1.3. Cost of natural disasters (Source: NOAA National Centers for Environmental Information, "Billion-Dollar Weather and Climate Disasters: Time Series," 2020, https://www.ncdc.noaa.gov/billions/time-series)

last five years being the hottest on record and the most intense storms earning names like "cyclone bombs." The EPA noted that certain populations may be especially vulnerable to climate impacts, including people living in poverty, people who are elderly, people already in poor health, people with disabilities, people living alone, and Indigenous populations dependent on one or a few natural resources. Anyone in the business of affordable housing knows that the populations we serve are feeling these impacts first and worse.

The Gray Housing Sector Is Itself Fragmented

Although some environmentalists might offer that the greenest house is the one that is not built, we do not have that option. We are short at least 7.2 million homes for those most in need and millions more are needed to make housing affordable for everyone, including those in the middle of the market. We are losing ground daily as existing affordable housing is bought by owners who rehabilitate it and then make it available to persons with incomes well above the average, causing displacement. The question is not whether to build or not to build. The question or problem to solve is how to meet our housing crisis while meeting other needs in our communities and making progress to mitigate the looming threat of climate change. We can do better. We must do better—and the ambition of our solution must be commensurate with the scale of the problem.

We cannot afford attempts to solve this crisis in our silos and separate sectors. The Energy Information Administration identified the industrial and transportation sectors in 2018 as the largest energy consumers and contributors to carbon emissions.[31] Because building consumption was distributed over various sectors, their impact was obfuscated. The same can be said about housing. Its benefits are distributed over various sectors that extend well beyond an individual household or person and certainly beyond any single public agency.

Federal Programs

There is no integrative approach or federal commitment to guaranteeing housing affordable to all. Because we think of housing, health care, education, transportation, jobs, and the environment in silos, we feel compelled to choose which issue we want politicians to focus on. The cost to meet the amount of housing needed in our country appears then to be in direct competition with our need to pay for the increasing costs of health care. The issue of meeting our housing needs vies for the attention of the American public as the need for family-sustaining jobs remains front and center for every politician.

This is evident at the federal level, where many departments have resources that can be allocated toward affordable housing beyond the obvious one of Housing and Urban Development. The EPA manages an important voluntary program, Energy Star, to help consumers save money through energy efficiency. The Department of Health and Human Services offers a critical safety net for home energy bills as well as repairs to weatherize properties that will lead to lower utility payments and greater comfort through its Low-Income Home Energy Assistance Program (LIHEAP). The Internal Revenue Service administers the Low-Income Housing Tax Credit, which is arguably the largest vehicle for financing multifamily affordable housing. The funding and capital provided through these programs are vastly underresourced, making them highly competitive to access. Stakeholders have to dance around agencies at every level of government and be savvy enough to know which opportunities exist and how to access them. Spreading all the resources around gives the illusion that all are doing their part to address the housing crisis and provides them enough latitude for finger-pointing and shifting the blame among entities.

The Development Process

Gray housing development operates in a vacuum measured by units constructed, available, sold, and rented. Gray housing developers do not recognize that housing is a fulcrum between the health and well-being of people and communities and the health of our planet. Generally, a developer purchases a plot of land wherever it is available, architectural designs are rendered or purchased and not always specific for that plot, engineering plans are finalized, a cost to build the housing development is determined, and the housing is financed and constructed. Once it is completed the developer may sell the housing, or retain ownership and lease it, or employ a third party to manage the property and lease the housing. The new owner or management company inherits the design and mechanical systems, as do the renters, who must pay the gas, electric, and water bills of the individual housing units. Everything is segmented into bits and pieces. The developer expects that if they build it, "they will come," regardless of how the individual person or family must orient other facets of their lives to fit the housing scenario.

The most critical players in the housing ecosystem are the residents of the community where the developments are or are proposed to be built. Community residents have the power of the vote when it comes to the overall policy direction for their community through the leaders that they elect. Those leaders then influence the type of development and growth that occurs and the budgets for the community's various offices and agencies. Residents also have the power of influence when it comes to local land use decisions that require public input or bonds to be issued to pay for housing development. This is partly why building housing in existing communities is more difficult than in the hinterlands, where there are no neighbors—but then there is also no connection to a community fabric. When community residents try to protect their own short-term vested interests, they can block or increase the costs for developers to build housing where it is most needed, which is in communities with opportunities for employment, quality public schools, grocery stores, and access to nature.

There are many great acronyms to describe the various stances community residents take on issues around development from NIMBY (Not In My Back Yard) to BANANA (Build Absolutely Nothing Anywhere Near Anything) and YIMBY (Yes In My Back Yard). Community groups can advance housing for all or they can choose silos and limit holistic development so that only people like themselves can live in their community. This can fragment the process of development by pitting interest groups against one another rather than finding solidarity in delivering housing for everyone.

Conclusion

We started this chapter from the perspective of housing as a broken system, and throughout this chapter, we reviewed the myriad ways that such brokenness is manifest across our communities. If we do not fix what appears to be so broken in housing, it will have lasting and potentially irreversible negative effects on the world we live in. The effects are seen in many areas because our systems of housing, education, employment, health, transportation, and land use, to name a few, are connected and must be considered as such. Housing is more than simply a real estate commodity. It is a platform from which to increase community cohesion, improve health outcomes, advance educational attainment, reduce carbon emissions, and enable community resilience. This is not the time to build more housing like we have always done, because that way of thinking landed us where we now find ourselves, with a growing wealth gap; a mismatch of housing type and housing need; half a million people living in public or abandoned spaces, not houses; and a planet that is heating up because of human-induced carbon emissions.

A home we can afford to rent does not mean much if we cannot afford the commute to our job. A home we can afford to rent does not mean much if it means living next to an incinerator or toxic waste facility that causes disease and cuts short our life span. A home we can afford to rent does not mean much if we cannot afford to turn on the air conditioning when temperatures climb. A home we can afford to rent is not really affordable if it is going to flood during an intense rain or if the roof is going to blow off during a storm. A home we can afford to rent does not mean much if there is no place nearby for our kids to play, for us to purchase fresh food, or for us to feel safe and recognized as equal to everyone else living in the surrounding community. And we cannot make housing affordable in the short run in ways that leave other communities worse off in the long run. A home is one ingredient in the overall recipe for opportunity. I argue that it is among the most important and basic of ingredients, along with clean air, clean water, and nutritious food. These are basic human rights, they are the foundations of any civil society, and they are recognized in the United Nations Sustainable Development Goals as contributing to a better and more sustainable future for us all.

CHAPTER 2

The Promise of Green

Action on behalf of life transforms. Because the relationship between self and the world is reciprocal, it is not a question of first getting enlightened or saved and then acting. As we work to heal the earth, the earth heals us.

—Robin Wall Kimmerer

Tracey was not looking for a green home, but simply one that she could afford to rent. When Tracey moved into the Guadalupe-Saldana development, she had a young child and was attending nursing school at night and working during the day. Guadalupe-Saldana is a net-zero subdivision, which means it produces as much energy as it uses, and it is located on a former brownfield site in east Austin. It was formed as a nonprofit organization governed by a board of residents and public representatives to provide lasting community assets and shared equity homeownership opportunities for families, referred to as a community land trust.[1] It was the first such community land trust with new construction in Texas and is a development project of the Guadalupe Neighborhood Development Corporation providing a total of 125 homes.

Admittedly, Tracey did not know everything that had gone into making her new home green, although she had noticed her apartment had solar panels on the roof to generate electricity and to heat water, and she had read news stories heralding its green qualities. Once she moved in, the green benefits materialized for her in two ways, lower electricity bills and better health. She immediately noticed that her electricity bill, which had been about $100 per month in her previous apartment, was now closer to $25 per month. Additionally, she noticed that her son's asthma was triggered much less often, which she attributed to not having carpet, but probably was further reduced because of the proper ventilation and air flow throughout the living space. Without allergies he slept much better, which allowed Tracey to get a full night's sleep, and her rested son started performing better at school. Tracey shared how thoughtful the architecture was and that it made her feel important to live in a home that had been designed with care and attention to detail. She was enjoying the amount of sunlight flooding into the space, and how easy it was to clean the kitchen countertops and floors because they were not cheaply tiled with uneven surfaces. These are all attributes of a green home. However, when you are primarily focused on finding access to a home with a monthly rent you can afford to pay, you do not always ask about the utility costs or about the quality of the indoor air or the durability of the finishes. Because green housing means you focus on the life cycle of the development and the people occupying it and surrounded by it, these become important details.

The benefits that accrued directly to Tracey are shared by the community, the schools, the larger region, and the country. Without Guadalupe-Saldana the neighborhood's brownfield site would have continued to be a blight on the surrounding area, negatively affecting the soil, and water quality, and degrading a sense of wholeness and safety to those living nearby. We all benefit from developments like Guadalupe-Saldana because in aggregate we generate less electricity from power plants, our children can breathe more easily and perform better in school, and families have a stable foundation from which to live with greater security. These communities then interlock across the landscape of our nation to strengthen the fabric of our entire society.

In this chapter I will discuss the field of green building, how it applies to housing, and its role in meeting our housing crisis (including its origins), as well as where we are today, and what more needs to be done.

What Is Green Building?

Green building is a process that prioritizes the health and well-being of the person living or working in that space, as well as their connection to the surrounding community. Green building does not sacrifice the health of the planet or other communities to construct or rehabilitate a building.

There are different ways to measure what makes a building green. At a basic level, almost everyone would agree that it should have some measure of energy efficiency, but beyond that the definitions vary. The Environmental Protection Agency defines green building as "the use of approaches that create buildings and development that are environmentally responsible, and resource efficient throughout a building's life cycle, from site selection to demolition or reuse."[2]

The United States Green Building Council (USGBC) describes green building as "a holistic concept that starts with the understanding that the built environment can have profound effects, both positive and negative, on the natural environment, as well as the people who inhabit buildings every day. Green building is an effort to amplify the positive and mitigate the negative impacts of these effects throughout the entire lifecycle of a building."[3]

No collective consciousness existed in the United States at any sizeable scale to holistically define a green building beyond the connection to energy and resource conservation until the USGBC created the

Leadership in Energy and Environmental Design (LEED) system in 1998. It provides a methodology for rating a building across categories (i.e., location and transportation, sustainable sites, water efficiency, energy and atmosphere, materials and resources, indoor environmental quality, and innovation) by assigning points to the various criteria in each category. A building is LEED certified for earning an established number of points above a minimum threshold (the different rating systems are shown in Figure 2.1).

The major accomplishment of the USGBC is that it created a platform for green building. It brought disparate views about what makes a green building under one tent to establish LEED. Although not everyone remained under one tent, the USGBC put the ingredients in place to create a green building movement and a system that assisted in monetizing the value of green building. They created a common language through which interested parties could interact and advance their collective action.

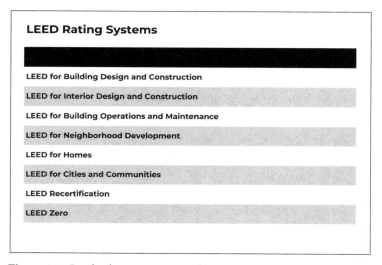

Figure 2.1. Leadership in Energy and Environmental Design (LEED) rating systems

The organization helps businesses that are building green to show that they are doing good. When a building is LEED certified, the owner receives a plaque to be placed prominently on a wall inside or outside of the building so that everyone will know it meets one of the LEED rating systems' requirements. LEED has evolved over time as developers, building owners and operators, and entire industries of consultants supporting them have demonstrated what is possible. It forced conversations about costs, transparency, materials, technologies, and values. What it did perhaps best of all was engage business leaders. It gave the environmentalists and the business community an opportunity to use this common language about green buildings. It provided an intersection between profit and planet where they could agree.

Over time, companies started competing, each wanting to be greener than the next. Doing so, reduced their operating costs and became good for their brand. Companies found that increased access to daylight and healthier indoor air quality boost the productivity of workers while reducing energy consumption.[4] Joseph Allen, an associate professor at the Harvard T.H. Chan School of Public Health and director of its Healthy Buildings Program, has extensively studied the impacts of indoor air quality on worker productivity. His work and that of others shows that "improving lighting, ventilation, and heating controls improves workers' performance, boosts their productivity, and even helps them sleep better at night. Developers and architects are starting to tout these benefits to potential tenants as a way to attract a higher caliber of employee—and get more work out of them."[5] The USGBC has helped make the green business case to top real estate developers and corporations. In 2020, the USGBC boasted over 94,000 LEED projects in over 165 countries and territories. In the age of COVID-19 it is imperative to improve the ventilation of buildings so that we are getting much more filtered fresh air into spaces instead of circulating stale indoor air. This can sometimes add cost to a building, but the health costs for not installing these features are now much more evident in the number of sick days, hospitalizations, and loss of life.[6]

Beyond USGBC

Although the USGBC gave us a language for talking about green buildings and an umbrella under which to learn and innovate and encourage one another, it is quite technical and practical. This encouraged others to innovate beyond the limits of LEED, which includes those who led the creation of the Living Building Challenge, run by the nonprofit International Living Futures Institute (ILFI). Rather than categories of requirements, the Living Building Challenge has seven "petals" that represent place, water, energy, health and happiness, materials, equity, and beauty.[7] In certain locations, what is required for a building to earn the Living Building certification is not even legal, available, or affordable, but it stretches us to consider what is possible. One of the early buildings to meet the Living Building Challenge is the Bullitt Center in Seattle, Washington, conceived of and owned by the Bullitt Foundation. The goal of the Bullitt Center as stated on its website, bullittcenter.org, is to drive change in the marketplace faster and farther by showing what is possible today. However, sometimes what is possible today is not available today. This is particularly true when it comes to materials. The team working on the Bullitt Center had a difficult time meeting the material petal because most manufacturers do not readily disclose the ingredients of their products. For example, the "makers of the liquid-applied building membrane used to seal the building and prevent heat loss, re-formulated their product when it was discovered that it contained chemicals from the 'red list' of banned substances under the rules of the Living Building Challenge™."[8] The hope is that, according to the ILFI website, these buildings will be "net-zero or net-positive energy, free of toxic chemicals, and lower their energy footprint many times below the generic commercial structure."

A green building is held to performance standards not associated with a gray building. It demands less energy and water and delivers healthy indoor environments while offering durable construction and interactions with its natural environment in ways that will be beneficial through landscaping and smart locations. However, the process of locating, designing, constructing, rehabilitating, operating, and maintaining a green building is the biggest area of differentiation from gray building.

The process of green building requires integrative design. It necessitates bringing a diverse team of stakeholders together, including people who live in the community and who will be living and working in the building. The design team must listen to the history of the place, the culture of the community, the conditions of the natural environment, and the needs of the people. Simply put green building uses the best of what we know about building science, new technologies, human-centered design, construction methods, healthy materials, environmental health, and climate change to deliver the type of building that will perform maximally for the conditions set forth through the integrative design process. Through the process of green building we can evolve how we deliver and rehabilitate housing.

Most of us, much like Tracey, are simply looking for a home we can afford in a place where we want to live. We are likely not looking for a LEED plaque on the wall but a rent payment or mortgage we can pay each month without forgoing all other basic necessities. However, the benefits of green building can make all of our housing more affordable, and we must commit to ensuring that housing for persons working for industries that pay the lowest wages is constructed and rehabilitated with the most effective green practices in place. These practices will ensure that housing remains affordable to operate and maintain and enhances the health of the people living inside and in the communities where the materials and energy are produced for that housing. We all benefit from cleaner air, water, and soil and a society where we all have a place to live that we can afford.

Why Green Matters for Housing

Green is an important strategy for meeting our housing crisis because it recognizes the complexity of that challenge and its interdependence with almost every other sector of society. Continuing to provide gray housing does nothing to provide a long-term solution to ensuring everyone has access to housing they can afford. Pushing people into housing that is affordable because we used cheap nondurable materials or we did nothing to keep utility costs low or to connect the housing to other necessities of life is not helpful in the long run. Green affordable housing recognizes the interconnectedness of our homes and our ability to flourish and sustain and nourish the planet on which we depend.

Today's green affordable housing movement is in part an attempt to align a community's need for access to affordable housing with the need for clean air, water, and soil. It is a way of thinking about housing that delivers high-performing spaces for people and uses methods and materials beneficial to the health and well-being of people and the planet. Green housing is not about installing "green bling," which is the act of installing materials and systems into housing because we think they are innovative or because we can, taking no serious consideration for how the materials or systems will operate together. Green is about holistically integrating proven methods and materials that will benefit the owner, the resident, the surrounding community, and the planet. Ultimately, green housing prioritizes the need to ensure that housing costs are affordable over the long term and from month to month.

Green affordable housing is about thinking differently to consider why we are making decisions, and for whom, and what we expect will be the outcomes of those decisions. We expect a green home to uplift the people living in it and regenerate the community in which it exists. That might seem like a lot to put on the table when we have a housing crisis and need to meet the demand quickly, but if we only provide shelter without considering the long-term implications of that housing, then we are missing an opportunity to do so much more than put up four walls and a roof. That opportunity involves meeting people's housing needs and their human needs while supporting a healthy planet and enabling a just and equitable transition to a green future for all.

The Green Communities Criteria

The standards for what we refer to as green affordable housing today, the Green Communities Criteria, were released by the Enterprise Community Partners' Green Communities Initiative in 2005. As a national community development intermediary, Enterprise Community Partners (hereafter Enterprise), seeks to deliver the capital, develop the programs, and advocate for the policies needed to create and preserve well-designed homes that people can afford in inclusive and connected communities full of access to opportunity. Enterprise was the originator of the idea to bring the benefits of green building to the affordable housing sector, and I was the staff person dedicated to bring this idea to life, working closely with a team from across the organization, including senior leaders focused on finance, policy, communications, and local housing solutions. I led the effort to create the Criteria in partnership with the Natural Resources Defense Council to ensure a rigorous environmental standard, and I collaborated closely with organizations that advised the process, including the American Institute of Architects, the American Planning Association, the National Center for Healthy Housing, Southface Institute, the Enterprise Rose Fellows,[9] and nonprofit community development corporations and staff from all business units across Enterprise.

We launched the Green Communities Initiative (chapter 3 provides more details on the development of the program) in 2004 because we thought bringing the benefits of green building to the affordable housing sector was imperative for two reasons: (1) it could contribute to making housing more affordable to own, operate, and maintain, and

(2) the affordable housing industry had a role to play in reducing carbon emissions and preparing communities for the impacts of climate change.

I designed the Green Communities Criteria (the Criteria), in close collaboration with all of the entities listed above, to be much like LEED but specific to affordable housing by emphasizing the use of proven methods and materials to drive down operating costs and improve health. I had the great fortune of learning from Enterprise staff who had developed a deep understanding of these proven methods and materials because of their experience working across the country over decades with community-based organizations to develop and rehabilitate affordable housing. This emphasis on health remains one of the Criteria's primary differentiating features in addition to focusing on cost-effectiveness. The building practices and materials included in the Criteria minimize moisture, provide proper ventilation, prevent pest infestations, and avoid toxic chemical exposure. To comply with the Green Communities Criteria, homes must be built in walkable neighborhoods that facilitate a healthy lifestyle incorporating physical activity. They are more energy efficient than gray housing, cutting utility bills by hundreds of dollars a year by providing Energy Star appliances and fixtures and increasing individual control of electricity. The Green Communities Criteria require that homes are built near mass transit or within walking distance to schools, jobs, and services, increasing opportunities and reducing transportation costs for families. Increased energy efficiency and proximity to transit will benefit families, communities, and the environment. Housing that meets the Green Communities Criteria has water-conserving fixtures and appliances. Low-cost site maintenance techniques are used to minimize erosion and manage stormwater on-site, and homes are built at appropriate densities, away from environmentally sensitive areas.

The Green Communities Criteria have categories that broadly define the components of green affordable housing. The main difference is that the Green Communities Criteria are hyperfocused on cost, health, and the requirement to engage in an integrative design process. The other components are site, location, and neighborhood fabric; site improvements; water conservation; energy efficiency; materials beneficial to the environment; healthy living environment; and operations, maintenance, and resident engagement. The main goal of the Criteria is to transform the way communities think about, design, build, and rehabilitate affordable homes. The following sections describe the rationale behind the eight components. The online Green Communities Criteria manual provides technical guidance on how to meet the Criteria.[10]

The Integrative Design Process

More than any appliance, material, or energy system, the defining feature of green building is the integrative design process. William (Bill) McDonough and Michael Braungart helped launch the green building movement with the 2001 film *The Next Industrial Revolution*, and their 2002 book *Cradle to Cradle*. The film was produced by Shelley Norhaim for Earthome Productions and was funded in part by the Rouse Family Foundation (Jim and Patty Rouse were the cofounders of Enterprise). Bill is a gifted orator who brings the technical aspects of green building to life by comparing a green building to the functions of a tree. One such quote, "Design is the first signal of intent," places accountability for outcomes on the designer of the system or building. So often we place it on the user. For example, society quickly assigns fault to the person who has been evicted, rather than examining our social systems for design flaws that prevented the person from paying the rent on time.

The integrative design process incorporates sustainability up front, uses a holistic and total-systems approach to the development process, and promotes good health and livability throughout the building's life cycle. The goal is to establish a written commitment that informs the project's objectives throughout the building's life cycle. Sustainable building strategies are considered from the moment the developer initiates the project.

The integrative design process embodies a fundamental principle about equity. A team that embraces an integrative design approach signals to all involved that they are important, they have a voice, and they have value to add to the process of creating a generative space where people will live. The integrative design process allows everyone to get on the same page and remember that they are involved in the sacred work of designing a place that someone will call "home." Everyone should understand the history of the land and the story of the place, and listen to the community, the residents, and the people who will be caring for the property over its lifetime. This is an opportunity to strengthen the fabric of that community. An integrative design process is not limited to new construction. We must increase the amount of new housing we are delivering while preserving, maintaining, and retrofitting existing affordable housing with the same care and attention we spend on new construction.

As Cindy Holler, former president of Mercy Housing Lakefront, said, "All of us deserve to live in beautiful spaces and healthy environments with beautiful views." Because beauty lies in the eyes of the beholder, we must listen to the people who will be living in the housing. The integrative design process sets the signal of intent for the entire development. When housing agencies were hesitant to adopt the full Green Communities Criteria, I asked them to start by, at a minimum, requiring an integrative design process to encourage developers and owners to explore what might be possible for the housing development. This allowed the design team to lead a process with all stakeholders about the history and context of the neighborhood, what the current and future residents wanted, and how to improve the overall performance of the housing for the people and the planet. Often the simple exercise of initiating an integrative design process to consider the possibilities made it evident that the full Criteria could actually be met. It often led to additional aspects of the development that would never have been considered otherwise.

One of the early supporters of the Green Communities Criteria, Jane Carbone, the director of real estate at Homeowners Rehab, and Rebecca Schofield, a project manager, were asked whether they experienced any surprising outcomes regarding engaging in an integrative design process. They shared that they "had a lot of hidden assumptions about what was possible . . . the ground floor design is a perfect example. We saw the plan as locked-in due to extensive requirements for internal bike parking and the space needed for handicap accessibility. We saw these demands as mutually exclusive to a community space on the first floor and we were looking at how to increase use of a third-floor community room."[11] Ultimately, Jane and Rebecca and their team responded to the input and located the space on the first floor where the residents were more likely to use it. They explored how to make exterior gathering spaces work not just for the residents but also for the surrounding community by making the community room available for community events. Because the integrative design process involves listening to the community, to everyone involved in the project, including the operations and maintenance staff, the input sometimes challenges the assumptions of the designers and results in changes that otherwise would not have been considered. It is about considering the development as a whole in the service of the people who live or will be living there and the community in which it exists.

The Arroyo Village, a magnificent development in Denver, was designed to meet the Green Communities Criteria while centering residents in every design decision. Arroyo Village includes a shelter for the unhoused, 35 apartments of permanently supportive housing, and 95 apartments of workforce housing. You enter into a community space instead of immediately encountering a security desk. There is a line of sight through the apartments no matter where you stand. This is a safety feature that the residents asked for after experiencing nights and days living in homes of their own making on the streets of Denver. Residents also asked for shelves and countertops to have a secure place for their belongings, many of whom had never had such a place. Human-centered design puts people at the center. I led the creation of the Green Communities Criteria to evolve and foster this combination of human-centered, community-centered, and planet-centered design.

Site, Location, and Neighborhood Fabric
A defining aspect of green affordable housing is that it is located in neighborhoods with existing infrastructure, such as sidewalks, quality green spaces and parks, and transportation options. It is also important that the neighborhood has services to meet the needs of the community. However, at the same time, we know that not all communities have such infrastructure or services, so it requires partnering with others to improve the existing neighborhood in ways the community wants. It is a both/and strategy of improving those neighborhoods without existing infrastructure and services while delivering and preserving green affordable housing in neighborhoods that already have it.

Proximity to services is critical, as we have seen during major weather events, when gasoline and public transportation options may be unavailable—and more recently, during the COVID-19 pandemic, when the requirement for physical distancing from others made public transportation less desirable. We witnessed improvements in outdoor air quality as a result of fewer people driving, but this came at a considerable cost to humanity, with over 200,000 lives lost in the United States alone. More resource-efficient development patterns encourage healthier lifestyles that allow for walking and biking and options for those with physical mobility challenges while reducing carbon emissions.

An increasingly important service we must provide access to with every green home is the internet. All homes, regardless of income or location, must have access to internet connectivity for remote educational opportunities, telemedicine, managing energy consumption, and critical services now offered online. To flatten the curve of the COVID-19 pandemic, we witnessed the closing of schools and the need for at least 50 million public school students in grades kindergarten through 12 to learn remotely from home; however, the digital divide will prevent many from accessing the education they need to stay on track. This is a problem across the entire country; no state is immune. Furthermore, up to 400,000 teachers cannot teach because of a lack of internet access.[12] Access to the internet is not a luxury; it is as critical as other basic needs like electricity and water. Not closing the divide is exacerbating inequities as we operate in an age where it is a necessity for functioning in our digital society.

The location of the development must be considered early on to naturally reduce energy demands for heating and cooling the buildings. This is referred to as passive solar heating and cooling. Even if it will eventually be renewably sourced through solar, wind, or geothermal applications, the more we can use nature to heat and cool our buildings without technology of any kind, the better—the more elegant too. There is something beautiful about interacting in a building that has been intentionally designed to be in harmony with nature and its surrounding environment.

Site Improvements

The basic question around site improvements is to ask how development can contribute to the health of the site. Most affordable housing sites have less to work with when it comes to site improvements because so many of the properties are on tight infill parcels of land. However, even on tight lots that have previously been developed, much can be done to introduce access to natural spaces or install cisterns underground that can capture rainwater, or water that has been used in the building, so it can be treated and reused. In the early days of green building, most of the attention on-site improvements had to do with planting only native species. Although that remains a concern because nonnative species can threaten the health of the ecosystem, managing stormwater on site has gained much more attention. This was a concern of an early development, Oleson Woods (Figure 2.2), that was designed to meet the Green Communities Criteria and that acquired a site in Tigard, the fast-growing suburb of Portland, Oregon. Here the neighbors were not too keen on allowing affordable housing to be constructed and did everything they could to rally against the developer, Community Partners for Affordable Housing, and keep the project from moving forward. Fortunately they were unsuccessful because now Oleson Woods provides 32 beautiful apartments at low rents with transit access and close proximity to employment centers. Community Partners for Affordable Housing was also able to preserve, protect, and enhance an existing wetland on the property that naturally manages stormwater but had been neglected and was so overrun with debris and waste that it was no longer functioning. The

Figure 2.2. Oleson Woods (Photo courtesy of Enterprise)

wetland began thriving and is now verdant. It has become a community attraction, providing interaction with nature for the residents and bringing surrounding seniors to the on-site community center to interact with the children in exploring the wetlands and learning of its many benefits.

The site improvements are critical because they can assist in bringing dignity into the development. What are those improvements that signal to the community and to the people living there that they are welcome, that this is their home, and that they are safe and seen? Thinking through the ways to lessen our footprint on the planet through sustainable site improvements is only part of the consideration. Site improvements within green affordable housing go further to ensure that the housing supports the mission of the development, respects the people who will be living there, and strengthens the fabric of the neighborhood. This requires involving the people in making decisions about site improvements that they want and can take a vested interest in caring for over time. In the integrative design process mentioned earlier with Jane Carbone and Rebecca Schofield, they were planning to simply plant a strip of shrubbery between their building and the property line. However, as a result of conversations with residents and community members and a desire to better connect the property to the existing neighborhood, that strip became a walking path. This amenity could be used by both the residents and the surrounding community.

Water Conservation

Water is not always a renewable resource. There are many communities already experiencing extensive droughts and prolonged water shortages. In parts of this country, the cost of water is higher than the cost of energy. Water conservation and harvesting must be a main pillar of green development. It is one of the areas that can provide the quickest payback through water-conserving fixtures and appliances and easy-to-install rain barrels. These can be low-cost, high-return investments.

Rainwater and wastewater can be harvested and treated for beneficial uses, including drinking, washing, bathing, toilet flushing, and irrigation. Buildings can meet all of their water needs with captured rainwater and wastewater and ensure that no sewage or stormwater leaves a site. Jane Carbone led an effort at Homeowners Rehab to create an underground cistern at an early Green Communities development called Trolley Square to capture rainwater on site to store and reuse (Figure 2.3).

Energy Efficiency

Because the operators and property managers of green affordable housing are engaged in the integrative design process, they know the performance expectations for the property (this is rarely the case in gray housing projects). The ventilation system in a green housing project is sized appropriately for adequate air flow based on the energy efficiency goals, which might be to meet passive house standards through

Figure 2.3. Trolley Square (Photo courtesy of Enterprise / Lloyd Wolf)

construction and design or to achieve net-zero energy usage by supplying what energy is needed through renewable energy systems. Many green affordable housing developments are starting to install monitors so they know in real time how much energy and water the building is consuming, compared to what it should be consuming for that time of day and season. The property managers can then more quickly address issues before the utility bill arrives.

Green affordable housing considers not only how much energy it consumes on site but also how much energy the materials and products going into the building consume. This is referred to as embodied energy, which accounts for all the energy required to produce the inputs to the finished product, including transportation. When designing a building with energy in mind, architects of green affordable housing consider both operational and structural embodied energy because the two are intertwined. For example, extending the roof out beyond the edge of a building can shade windows and reduce cooling needs in hot climates, but making an overhang that is structurally sound can take a lot of energy-intensive material. These tradeoffs need to be considered.

Materials Beneficial to the Environment
Green affordable housing pays attention to where building materials come from and where they go. It supports a materials supply chain that is beneficial to the environment from the extraction of the materials and throughout their life cycle. Green affordable housing uses materials that reduce our consumption of natural resources. For example, a project may use cool roofing materials that lower the ambient air temperature by reflecting sunlight and absorbing less heat than a conventional roof. This reduces the amount of electricity needed to cool the building and reduces the stress on the surrounding plants and trees and the additional need for water.

Green affordable housing uses advanced framing techniques, which reduce the amount of wood used and result in less waste by cutting the wood only to usable lengths. Where possible, unused wood is reused elsewhere in the building. The use of salvaged wood, or sustainably harvested wood products certified by the Forest Stewardship Council, or not using wood at all, is part of what it means to be green. Less than 10 percent of old growth forest remains in the United States at a time when we desperately need more forests to sequester carbon and other carbon emissions. Green materials do not include composite wood products with formaldehyde-based binders.

The bottom line regarding materials in green affordable housing is to use only what is needed; know what is in the materials used; know how the materials will affect the workers and people exposed to manufacturing, installing, and cleaning them; and not use anything that cannot be reused or recycled. Like any good environmentalist, shop locally. The manufacturing of green building materials can be a source of economic growth.

Healthy Living Environment
We spend 90 percent of our time indoors, and for the elderly, caregivers, and small children, this often means we are spending 90 percent of our time in our homes. It is imperative that we construct and rehabilitate housing to be healthy. This means testing for and mitigating radon, which is the second leading cause of lung cancer in the United States, and eliminating exposure to lead while implementing known measures for improving the health of the workers and residents through healthy materials and proper ventilation. It may seem counter to our energy efficiency goals, but mechanical ventilation is essential to delivering a healthy living environment. The use of energy conserving fans and systems will keep additional energy consumption under control, and the payback in healthy residents is well worth the small trade-off.

Our homes contribute to our health by connecting us to our community. The former surgeon general of the United States Vivek Murthy found loneliness to be the most common pathology afflicting us. Designing our homes to allow light and spaces for residents to gather and connect improves our health. This can be achieved through both universal design and healing-centered design. The Center for Universal Design recognizes universal design as the design of products and environments to be usable by all people, to the greatest extent possible, without the need for adaptation or specialized design. Healing-centered design ensures that residents are not reminded of past trauma and are able to move toward healing. Both approaches enhance and support social cohesion and interpersonal connection.

Operations, Maintenance, and Resident Engagement

The bookend to an integrative design process is the operations and maintenance of the property. It is like taking a multiple-choice test and finally seeing the answers to know how well you did. The operations and maintenance should be much more predictable in green affordable housing because of the intentional way the development was designed and the clear performance targets that were set. There are humans involved so it will not be entirely without surprises, but generally speaking the performance of green affordable housing should be far less risky than that of its gray counterpart.

How often does anyone examine their utility bills? We see the amount we owe and we do our best to pay it because we do not want our electricity, gas, or water to be shut off. Do you know what to expect in terms of the gallons of water or kilowatts of electricity that your home is supposed to be using? Sometimes making sure we are billed the correct price for a gallon of water or kilowatt of electricity can result in a lower bill. This is where the operations and maintenance staff are critical to the housing system, but they are rarely the people you find at an affordable housing conference. Their experience is invaluable in helping to design a development that will not only lower costs but support the long-term health of the property and the residents while not contributing to the poor health of staff, who can be exposed to toxic classes of chemicals in the cleaning materials and in the adhesives and products used to repair fixtures, finishes, and appliances. They are also more likely than most of the other players to have direct interaction with residents to know what else is going on in the building that might be detracting from its proper functioning.

Launching the Green Communities Initiative

Enterprise, NRDC, and our partners, launched the Green Communities Initiative at a time when many were grappling with what part they could play to mitigate climate change. It provided an umbrella for people's interest in the environment, public health, renewable energy, smart growth, and community development. Perhaps even more impactful was that it signaled to the affordable housing community that it was time to embrace a different approach and to deliver 21st-century solutions to our housing crisis. NRDC was not as involved in the implementation of the Green Communities Initiative but did remain engaged to ensure that developments earning the title of having met the Criteria did in fact do so. Enterprise took the lead in demonstrating that we could preserve affordability and address building practices that aligned with health and environmental sustainability. This was no longer a question of choosing to do one or the other—it was time to do both green and affordable housing. Enterprise charted the path to show that both are possible and how both are mutually reinforcing.

The Enterprise Green Communities[SM] Initiative was initially a five-year, $555 million program of tax credit equity, investments, loans, grants, consulting services, and training intended to finance and promote the development of green affordable units for rent and homes for sale that provide significant health, environmental, and economic benefits to families who are earning low wages and are underinvested in communities. The philanthropic support came from the Bank of America Foundation, the Citi Foundation, the Home Depot Foundation, the Kendeda Fund, and the Kresge Foundation. Their support was critical to the overall effort because it allowed us to use grant funding as an incentive for those nonprofit developers who had a steep learning curve, needed a carrot to convince their boards, or had a specific challenge for which they needed additional funds. Internally it was extremely helpful to have outside entities to whom we were responsible for delivering on our goals. With their support, I was able to push back on internal pressures to weaken the Criteria or shrink our ambition because we had agreements with our funders to demonstrate impact. One funder provided a large grant that we used as a guarantee against loans we wanted to make that carried potential risk if the energy and water savings were not realized. Taking financial risk off the table unlocked other forms of capital that allowed owners to engage in retrofitting their property to higher performance standards than merely replacing systems with what had always been used. For example, they could purchase appliances known to use less energy because they would be able to pay back the cost of the loan through savings on utility bills. We could draw on the grant as a loan guarantee if they did not. Now financial institutions offer this type of financing, but in the early days we were able to provide proof of concept by using grant funds.

Bart Harvey, CEO of Enterprise at the time and one of the primary visionaries behind the Enterprise Green Communities^SM Initiative stated at the launch event that "too many Americans live in unhealthy, inefficient, and poorly sited housing that hinders them from reaching their full potential. Enterprise and the Natural Resources Defense Council have forged an unprecedented alliance of housing, health, and environmental organizations to ensure that smarter, healthier homes are available to Americans with limited incomes." Dr. Megan Sandel, a nationally recognized expert on housing's impact on children's health at the Boston

University School of Medicine, explained that "one of green housing's major selling points is that it means healthier indoor environments. For many families, asthma, injuries, and lead poisoning are just symptoms of the underlying problem. Inadequate housing is the real disease. Safe, decent, affordable housing is the best preventive medicine low-income families can get." The focus on health remains one of Green Communities differentiating features from other green building programs, ones more focused on energy conservation.

The press event was a critical ingredient in the systems change we were seeking. We had to signal to the world, in a very official capacity what we were setting out to do. We had to draw a line in the sand; declare a big, audacious goal; assemble legitimate stakeholders endorsing the goal; and then send that message far and wide. It was the press release that rippled across the country and reverberated with calls from mayors, governors, housing officials, and nonprofit leaders wanting to participate. It was about pulling the lever of communication to send waves out across the country signaling our intentions to bring the benefits of green building to people who might gain the most. It catalyzed the Minnesota (which was the first to do so), San Francisco, Massachusetts, Michigan, and Florida Green Communities Initiatives that pulled resources and incentives to demonstrate how developers and owners could meet the Criteria in those states.

I decided not to franchise Green Communities Initiatives in every state because of limited staffing resources. I did enable these five during the first few years of launching our efforts because of the strong interest from those states and because each was solving for a different challenge that could be helpful in demonstrating that the Criteria could be applied across the entire housing sector. Minnesota needed to make sure the Criteria worked in rural settings. San Francisco was focused on dense infill redevelopment and containing costs while serving formerly homeless populations. Massachusetts was eager to leverage clean energy partners.

Michigan was hoping to strengthen its overall housing industry, and Florida wanted to explore whether green methods could lower operating costs for homeowners while reducing sprawl development. Each of these efforts required working groups made up of Enterprise staff and stakeholders from the housing finance agencies, community development financial institutions, housing offices, developers, owners, mayors and governors, staff, technical assistance providers, and residents.

The state-based work accelerated Enterprise's understanding of challenges to meeting the Criteria and enabled those states to learn together how best to address challenges and opportunities and to move the entire housing sector along. We applied less formal collaborative approaches in most states by involving all parts of the affordable housing ecosystem in that geography, tailoring the technical assistance to local nuances, and delivering regional workshops so everyone could learn together. This model forced us to grow the field of technical assistance providers because we experienced how uneven that playing field was from market to market. If development teams had access to technical assistance providers that understood building science and affordable housing, then we noticed easier adoption of the Criteria at lower costs with higher performance. I eventually led an effort to create a technical assistance provider network staffed by the Enterprise Green Communities[SM] team because of how important these entities were to the evolution of the Criteria and to their local implementation. Individual providers became key advocates for affordable housing and often supported efforts to inform the development of policy at the federal and local levels related to making all affordable housing green.

As already described in this chapter, the Green Communities Criteria promote integrative design; site, location, and neighborhood fabric; site improvement; water conservation; energy efficiency; materials beneficial to the environment; healthy living environments; and operations, maintenance, and resident engagement. Rather than measuring affordable housing by the number of "units" affordable to a certain income bracket, the Green Communities Criteria set a new benchmark for measurement. It was intended to go beyond first costs and to consider how the housing would perform for the resident, the owner and operator, the community, and ultimately the planet. The Green Communities Criteria were not a simple checklist; they were intended to inspire a new way of thinking. It was this bold push that caught the imagination of the entire affordable housing sector. Suddenly, there was new energy and excitement about what was possible, and we were soon able to share evidence that the Criteria could be met for little to no additional cost and that doing so resulted in better health outcomes and long-term operating savings for the residents and owners. This resulted in adoption of the Criteria by an increasing number of housing agencies. All of this work was made possible because of the philanthropic support provided to Enterprise—a terrific example of how philanthropy can enable innovation within the nonprofit sector that is then adopted by the public sector.

The origin of the idea for the Green Communities Initiative is not directly attributed to any single person. Before officially launching the initiative Bart had been influenced heavily by Greg Kats, Diane Miller, Jonathan Rose, and Stockton Williams to consider how Enterprise might join this budding green building movement. Jonathan was on the board of Enterprise and already engaged in green affordable housing through his development company, the Jonathan Rose Companies. Greg was among the early founders of the USGBC and active in clean energy investments. Stockton Williams saw the opportunity to raise new grant support and embed green into policies and financing. Additionally, Diane was actively funding green building efforts through her foundation, the Blue Moon Fund. Having a private developer, an economist, a policy expert, and a philanthropist prodding Bart to go green proved to be the necessary reassurance in compelling him to go all in.

There were robust discussions among Enterprise's trusted advisers about what constituted "green." How would it be defined for new versus existing construction, and for rural versus urban and suburban locations? Fortunately, Enterprise created a robust framework that could work for all housing because in the ongoing efforts to have federal agencies, states, and cities adopt the Green Communities Criteria, the first question is always, "Well, what exactly is 'green,' and how much does it cost?" If that framework had to be re-created for every situation and every jurisdiction, then the challenges of widespread adoption of a single framework would have been insurmountable. Being able to say, "Here's a definition of green affordable housing called the Green Communities Criteria, which has been tested and evaluated for cost and benefits" made the program possible. The Criteria are designed to work in any climate, for any type of housing, in any community. Having a flexible-enough framework driving toward measurable outcomes allowed us to pivot from needing to agree on a definition for "green affordable housing," which could take years to implement. Often it happened first through adopting the Criteria for a voluntary pilot in places like Minnesota and San Francisco, and then to wholesale adoption required to access housing programs and financing. In Minnesota the coordinated housing system supported by the Greater Minnesota Housing Fund and the Family Housing Fund provided technical assistance, funding, communications, and consulting support to assess every aspect of the housing system that would need to be in alignment to fully realize the benefits of green. In San Francisco it took the leadership of then mayor Gavin Newsom to unlock resources available through the city to demonstrate what was possible so that commercial owners and tenants were not the only ones benefiting from green.

There could have been a wave of green building standards crashing across the affordable housing sector. Part of what keeps housing costs low for developers is knowing what is expected of them by the entities providing the approvals and financing. Many developers work across state boundaries so it is helpful to have a basic framework that can be understood in any jurisdiction. This is why other programs like LEED have fared so well. Without clear expectations and with so many things to consider, most developers will default to what they did before. This is not a judgment; given the hoops, rules, and regulations that exist in this industry, it is amazing that anything gets built at all. The development of affordable housing is not for the faint of heart.

Development of the Criteria

A five-person team, including myself, drafted a framework for the Green Communities Criteria over a 48-hour retreat in the Timpanagos Mountains of Utah at the Sundance Resort.[13] A sacred place to consider the impact of our built environment on natural systems and an interesting one in which to contemplate the future of affordable housing. With help from Joanne Quinn at the City of Seattle Office of Housing, we obtained permission to essentially borrow SeaGreen, a guide for greening affordable housing in Seattle, which was based on a guide that the City of Portland had created, which was in turn based on Austin's green building standard. The goal was to ensure that the Green Communities Criteria relied on proven methods and practices, and that they did not promote any "green bling" items that provided no benefit to the resident, owner, community, or planet. The framework of the Criteria aimed to change the way we think about, locate, design, and build new construction and rehabilitate affordable housing so that it provides significant health, economic, and environmental benefits for residents earning low wages across the country.

We wanted the Green Communities Criteria to require that all involved understand that this was a different way of thinking and not about bells and whistles that could be applied to a building in some ad hoc fashion to accrue points. This was going to require intention and up-front commitment by the whole team involved, including those approving the designs, providing the financing, managing the properties, and serving the residents. Only through this approach could we guarantee that meeting the Criteria would deliver the benefits of green building that would lead to better housing with lower operating costs and greater health benefits for residents and for the planet.

We decided early on to ensure that the Green Communities Criteria would be as rigorous as the USGBC LEED rating system but with more mandatory items to take out the guesswork and ensure that we could measure the benefits from all housing developments that met the Criteria. This resulted in my needing to understand the LEED rating system, which involved a crash course in green real estate at MIT taught by Bill Browning, a founding member of the US Green Building Council's board of directors, a training workshop facilitated by Southface in Atlanta, and then becoming LEED accredited in August 2004. It was evident early on to me through these experiences that the Criteria were more focused on health than other green building programs. Over time, health has become one of the biggest selling points for green buildings, including LEED-rated ones. Commercial tenants were observing the increase in productivity of workers with access to daylight and proper ventilation that allowed for the right mix of fresh air into the building. We wanted people to experience this in their homes too.

I knew we must have a uniform standard across the country so that every person seeking affordable housing could have access to the same benefits. If we created a standard with only optional Criteria, then it would be luck of the draw whether or not someone living in New Orleans would receive the same benefits from living in homes meeting the Green Communities Criteria as someone living in Atlanta, El Paso, or Oakland.

When the Green Communities Criteria were first rolled out, I relied extensively on the practical knowledge of the Rose Fellows (particularly

Michael Gatto) and on Southface Institute (primarily Gray Kelly), with whom I later contracted to support trainings for Enterprise staff and eventually for community development corporations across the country. There was a renewed commitment by many to the mission of the work because they felt that this was something bigger than Enterprise.

The value of working with Southface was that it focused on building science. Other partners were more focused on policy, and many did not have experience with affordable housing or even residential construction. To be a change agent you need to have legitimate standing regarding the change you are seeking to make.[14] Southface, which focused on building science, and Enterprise, steeped in affordable housing experience, made terrific partners. Southface could vouch for the science behind the Criteria because they had put it into practice with their own green building standard, EarthCraft, and Enterprise could vouch that the Criteria were the right fit for affordable housing because they were willing to provide the funding and financing to make it happen and were committed to increasing the supply of affordable housing and would not do anything that would result in less housing production and rehabilitation.

Testing the Criteria

To be successful, the Green Communities Criteria must be attainable by all, not only the top tier of developers. The Criteria must not lead to additional expense, possibly resulting in a lower number of affordable rental or for-sale homes available to persons with low and very low incomes. This is why it was imperative to test the Criteria on a range of completed developments that were on the leading edge of delivering what they considered to be green housing. There were staff within Enterprise who wanted a short checklist of optional items that if met would lead to labeling a development green and providing more favorable financing. They were not in favor of a holistic approach that would require all criteria to be met. Therefore, the future of the Green Communities Initiative and the Criteria rested on results from testing the Criteria to see if meeting them was achievable.

I worked closely with Peter Werwath at Enterprise, a seasoned afford-able housing developer and technician, to choose five developments that Enterprise had provided financing to: Holland Apartments, Plaza Apart-ments, High Point, Denny Park, and Sanctuary Place. We sent the draft Criteria to a few of the projects that were under construction so they could share with us which items the development would be meeting. If they were not meeting a criterion, we asked them to explain why. If the reason was cost, we asked them how much each change was expected to cost above what they would have chosen to do. The strategy here was to understand if the method or material was actually driving up costs or if it was a perception of higher costs or possibly an unfamiliarity with a criterion that was causing a hurdle. Answers to these questions would allow us at Enterprise to reconsider what we were including in the Cri-teria and to provide clear guidance in a technical manual about how to implement each criterion in different climate zones, building types, and housing markets.

I visited the developments and interviewed project managers, archi-tects, and other professionals involved with the execution of these projects. By evaluating the responses, we determined whether some measure of the Criteria could be realized, especially with respect to their achievability for all housing construction types as well as any regional variability to which the Criteria might be subject.

Four of the developments were located in urban areas and one, Hol-land Apartments, was in the rural town of Danville, Illinois, with a pop-ulation of 35,000 people. Sanctuary Place was in Chicago. Two were in Seattle: High Point, and Denny Park Apartments. Plaza Apartments was in San Francisco. The four urban projects were managed by develop-ment teams that were very experienced and demonstrated a high level of awareness with respect to the issues addressed by the Criteria. Although the development team operating in a rural setting had some development experience, the project manager was less informed about the Criteria. In fact, he noted at the outset that he would benefit from technical support in initiating such a "green project." In addition, there were fewer resources locally available to him to support the project. Although the rural team's design professionals were aware of green building and engaged in managing several green features in the current project, they came to Danville from other towns where more green commercial construction was underway.

The level of support for green building in the area determined the likelihood that a jurisdiction would (or would not) adopt the Criteria into its housing programs. Levels of support might manifest in more enlightened codes, the availability of more money targeted for the environmental and sustainability aspects of the development, and the level of awareness not only of professionals but the population at large. Each of these attributes was well represented in all three of the West Coast projects and observable in the Chicago project. We did not make a formal effort to quantify the level of community sophistication around green building or its level of support, but maybe that should have been factored in. When a development team has to spend time educating its members on methods and materials, it can add to the predevelopment costs and may require more oversight during construction, which can also add costs.

Across all five projects, there were only a few times that a criterion was not met because it was deemed infeasible. In most cases it simply had not been considered. Four projects failed to meet the criterion requiring the use of wood certified by the Forest Stewardship Council. The developers said that the criterion was not yet supported by a mature supply stream and was not available at places such as the local Home Depot. This is something that has definitely changed since then in terms of wood products. The evolution of the green building field and the demand for certain products have shaped what the professionals are comfortable specifying in their projects. However, the availability of environmentally beneficial and healthy materials and their respective performance data remains limited. The other criterion involved reducing the thermal gradient differences between developed and undeveloped areas by using a vegetated roof on over 50 percent of the roof area or by using roofing with a minimum solar reflectance index (SRI) for the roof slope. Several developers said that spending money on this did not seem to make sense. That was understandable and a consideration for the future Criteria. If a criterion has a societal benefit but an incremental cost to the developer, should it be mandatory? This criterion supports the overall goal for a community to lower its heat island effect, which consists of varying temperatures within dense urban areas and the surrounding community due primarily to more paved surfaces and less vegetation. However, this criterion also provides a direct benefit to the owner or occupant because a cool roof can reduce the need for air conditioning, lower temperatures in the summer months in spaces that are not air conditioned, and reduce the amount of heat absorbed by the roof, which may extend its useful life.[15]

The Seattle developers noted that because the weather is temperate (many homes do not have air conditioning because it is not needed), concerns about the heat island effect were moot. However, the impacts of climate change are becoming more widespread, and Seattle experienced extreme heat in 2019. This criterion is gaining importance as a key resilient building technique in most parts of the country where the urban heat island effect is real and becoming life threatening because people with limited incomes do not have air conditioning, or cannot afford to keep it running, and may have no transportation options to escape the area when the heat becomes unbearable. This can be catastrophic for the elderly and for anyone with a cardiovascular or respiratory illness.

There were few consistent patterns in the differential costs noted in the interviews for meeting the Criteria. Four projects had noteworthy expenses for ventilation, and the cost per unit ranged between $75 and $350 over what they would have spent. Many of the Criteria were satisfied by all of the projects with little effort. Only Holland Apartments had difficulty with the "low volatile organic compound (VOC)" criterion, and in that case it was simply because the developer did not know that such a product existed.

It would not be the first time something like low- or no-VOC paint would be a stumbling block. When the National Housing Trust embarked on one of its first Green Communities developments that involved significantly renovating existing residential buildings, the contractors told the project managers that they could not afford the non-VOC paint. The National Housing Trust was an early adopter of green because they saw this as critical to their mission to preserve existing affordable housing. Besides not being able to afford the paint, this development was meeting all of the Criteria. The other projects we were testing were not faced with this cost differential, which we conveyed to the National Housing Trust, so the project manager called the manufacturer and verified that indeed there was no additional cost for non-VOC paint. This is an example of how having a network of projects connected by a trusted intermediary can be useful for sharing information, truth testing, and exchanging what is working and what is not.

The back testing gave me and others within Enterprise, including Bart, the confidence needed to defend the comprehensive approach it was taking with the Green Communities Criteria. Prior to the back testing it had all been an academic exercise. It is a compelling case on paper, with the need to reduce carbon emissions, to waste less construction material, to avoid using toxic chemicals, and to build within the existing fabric of a community. Without the human stories of why the green aspects matter to people, the Green Communities Criteria could be waived or easily dismissed as an obstruction to delivering the number of affordable homes needed. However, when you speak to the people involved in the developments, it becomes obvious that green building is more than a compliance checklist. Most notable was a conversation we held with the developer of Sanctuary Place about the women who had formerly been incarcerated or had been unhoused for years, and as a result, had compromised immune systems. It was important to ensure that the materials used inside these homes were not further harming their immune systems by off-gassing toxic chemicals.

Cheaper building materials often contain toxic chemicals, such as glues and adhesives that are used in their fabrication. People living in housing where deteriorating materials are off-gassing are more likely to be susceptible to a virus that attacks the respiratory system. With the requests for everyone to shelter in place during the COVID-19 pandemic, we have yet another reason to ensure that our homes do not include toxic materials.

Children in the High Point community in Seattle placed signs on trees noting the trees' financial value to help preserve the many old growth trees on site, which help filter particulate matter in the air resulting from industry and traffic. In many developments those trees would have been bulldozed to make siting houses easier. The intentionality of green building permeates more than a materials list. It allows for the community to preserve the culture and assets in their neighborhood because they are engaged in the integrative design process. The development is more than just shelter. It is about signaling to the community that they are important, and they deserve access to nature, healthy living environments, and living conditions that enable them to flourish.

High Point remains today a hallmark of excellent design. (The neighborhood received its name because it is at an elevation of 510 feet, thus being the highest point in Seattle.) The original High Point public housing project was built in 1942, when the US entered World War II, and an influx of workers moving to Seattle to work at Boeing and other war-related industries created an immediate need for housing. The Seattle Housing Authority (SHA) addressed this problem by rushing to complete over 700 units. After over 60 years of use the SHA applied for and received a HOPE VI[16] grant to assist in the redevelopment of High Point. It created 1,600 units of housing on a 120-acre site. Beyond the sheer number of homes affordable to a variety of low- and very-low-income households, the redevelopment stood out for me because of other notable features. The primary one being the new, transformative layout with access to public transportation, an extensive sidewalk network, and water amenities.

Rainwater typically collects all sorts of contaminated dirt and materials as it flows over impermeable surfaces. In the original High Point development, which had few permeable surfaces, the rainwater runoff flowed into nearby Longfellow Creek, one of Seattle's most important salmon streams. The new development was designed with a natural drainage system running through the site, one of the largest in the country, to enhance the creek and not endanger the health of the salmon. The system includes natural features such as vegetated swales, stormwater cascades, small wetlands, and a retention pond, all contiguous, running throughout the 34 blocks (Figure 2.4). Numerous pocket parks were

Figure 2.4. High Point (Photo courtesy of Enterprise / Neil Poulsen)

designed, and new services include a library, coffee shop, neighborhood center, and health and dental center.

The natural drainage system cost more than a traditional system. The City reimbursed the SHA for the extra costs. In contrast, the size of the pond was smaller than it would have been using a conventional system so it allowed SHA to build more housing.

Because all five developments met or could have met the Criteria, it became clear that what we were offering was an achievable integrative approach to delivering measurable health, economic benefits, and environmental benefits. This was true even for the Holland Apartments, a prominent 100-year-old building in Danville that was listed on the National Register of Historic Places. Additionally, 12 of the 46 units were reserved for individuals or families who had been unhoused or had disabilities. The remaining units were for persons with low and very low incomes. I recall feeling overwhelmed with the scope of the rehabilitation and the constraints that came with the building's being listed on the National Register of Historic Places, but I also understand the importance of reinvesting in those important structures. When I visited, it was utterly run down, vacant, and at risk for demolition. We are talking about more pigeon poop on the inside of a building than seemed possible. I will never forget it.

After walking through the structure I would not have guessed that it would be able to meet the Criteria, but meeting Thom Pollock changed everything. Thom was the executive director of Crosspoint Human Services, a large nonprofit agency established in 1980 to provide services to 1,800 clients with either developmental disabilities or mental illness. CHS was going to be the property manager for the Holland Apartments. Thom was a local preservationist and affordable housing advocate and had convinced his board of directors to purchase the building, setting in motion a plan to blend affordable housing, permanent supportive housing, historic restoration, and green development. On the green building end he was thinking about the use of geothermal heat pumps, which were taking off in Chicago. The technology uses the constant temperature of the earth to heat and cool a home. He probably was not thinking about the tension between being green and remaining true to its historic heritage. The two goals did not always align. Thom recalled having to "pull all the windows out, strip off all the lead paint, and refinish them, and put them back in their holes." This did not allow for replacing the inefficient windows with more durable and energy-efficient ones—a major policy flaw.

Another constraint probably not considered at length when the project was first conceived involved the requirement to use the same type of bricks currently on the façade of the building for the improvements. He had to search all over the region for what would be acceptable to the National Register of Historic Places. As a result of Thom's sheer determination, he found another demolition site and put some of the future residents to work cleaning those bricks to be used in the Holland Apartments. During construction, as the contractors were ripping up old carpeting, they found an ivory porcelain floor, which reduced the need to purchase any additional flooring because it provided a durable, healthy, and easy-to-clean surface far superior to anything the development team would have been able to purchase. The determination of one person within a development became the common element of these projects; one person can make all the difference.

With the back testing complete and the knowledge that cost-effective green affordable housing already existed, we were confident in our robust approach. The Green Communities Criteria were officially released in February 2005.

Conclusion

The Criteria have evolved since their release in 2005 through input from some of the nation's leading environmental, public health, green building, and affordable housing experts and practitioners and residents. This green building framework was the first (and remains the only) framework in the nation to address the unique needs of the entire affordable housing sector. At least 27 states now either require or incentivize the Criteria to be met. Without Enterprise's persistence to bring the benefits of green building into the affordable housing sector, it is fathomable that the affordable housing sector would have been left behind. Instead, the sector that has the most to gain from the benefits of green building is able to cost-effectively, rigorously, and holistically do so.

CHAPTER 3

Learning from the Green Communities Criteria

Green Communities significantly contributed to the advancement of green building practices in the field, primarily through development of the Criteria and extensive work on the policy front.

—Ross Strategic's evaluation of the
Enterprise Green Communities[SM] program

THE GREEN COMMUNITIES CRITERIA MAKE UP the most referenced sustainability program for affordable housing in the United States. They are a powerful platform influencing most of the construction and rehabilitation that are financed using public and private subsidies to make housing affordable to people earning incomes that are not commensurate with the cost of renting and owning housing in their own communities. As a member of Enterprise's Real Estate Leadership Council said in an interview with Ross Strategic, who were evaluating the Enterprise Green Communities[SM] Initiative, "had Green Communities not been launched, we are not sure whether there would be any green building standards for affordable housing in the US today."

Because of reports conducted and commissioned by Enterprise, it is known that green building methods and materials can deliver significant health, environmental, and economic benefits. The result is that people can better afford their housing and society can better deliver housing that is more affordable over the long term. People are able to live where they need to live because the housing is smartly located and connected to the neighborhood. People are healthier, including both those providing the materials used in the housing and those living and working in it. The planet is better off because of less pollution, conservation of resources, and a lower carbon footprint overall, and the systems involved in how communities develop and thrive are interconnected through the integrative design, planning, and development process.

Housing Is More Affordable to Operate
The costs and financial benefits of meeting the Criteria were evaluated in 2009, in a study looking at 27 projects, and again in 2012. The major finding in the 2009 study, completed in consultation with Performance Systems Development, was that meeting the Criteria on average costs $4.52 per square foot, or about a 2 percent upfront investment in total development costs while providing significant long-term operating cost savings. Energy and water conservation measures not only pay for themselves ($1,900) but also produce another $2,900 in projected lifetime savings per dwelling unit. From a strictly financial standpoint, the projected "lifetime" utility cost savings, averaging $4,851 per dwelling unit discounted to 2010 dollars, are sufficient to repay the average $4,524 per-unit cost of complying with the Criteria. In summary, estimated lifetime savings exceed the initial costs of incorporating the Green Communities Criteria into affordable housing.

In a 2012 study Davis Langdon, now an AECOM company (an infrastructure consulting firm), completed the analysis of 52 Green Communities properties, which included more rehabilitation developments than the first study. The average project analyzed in this study achieved a lifetime utility cost savings of $3,709 per dwelling unit, and the average project cost to comply with the Criteria was $3,546.

Both studies found that lifetime savings exceed the cost of integrating the Criteria into affordable housing, and both studies found it incredibly difficult to collect design and construction cost and utility consumption data.[1] Much like the first study, the energy and healthy living environment criteria drove most of the incremental costs, with high variability in actual versus predicted energy and water consumption levels. However, the cost of healthier building materials had decreased in the second study. As development teams completed more than one green development, they could minimize costs and maximize savings by applying lessons learned from one project to the next. The 2012 study revealed that the median cost of implementing the Criteria was $3,546 per unit. This represents a 1.85 percent increase to the total development cost for the project. The median cost to integrate only the energy and water Criteria was $1,139, and this returned $3,709 in predicted lifetime utility cost savings.

Over time we were able to show that properties that had met the Criteria were saving up to 40 percent more in energy costs than gray properties. The first Green Communities development, Denny Park, owned and operated by the Low Income Housing Institute (LIHI), saves as much as $200 per apartment compared to utility costs at other LIHI gray affordable housing properties. LIHI saved on its water and sewer bills at Denny Park too. Denny Park resident Ed Keys, who lives on a fixed income, realizes energy saving of about $200 a year. The American Council for an Energy-Efficient Economy reported that improving the energy efficiency of the homes of families with high energy burdens would significantly cut or eliminate the excess energy burden. For Black and Latinx households, 42 percent and 68 percent of the excess energy burden, respectively, would be eliminated by making their homes as efficient as the average home. Furthermore, energy efficiency in our homes results in lower demand at the source—for example, fossil fuel extraction. Because LIHI worked with residents at Denny Park to remind them about the importance of recycling, it further saves over $100 per apartment compared to LIHI's other gray properties. Recycling is not a mandatory criterion in the Criteria, but providing information about the building's green features with residents is required.

The major cost savings for green affordable housing are lower energy and water bills. These are probably the more obvious benefits, but the economic benefits extend beyond the entity responsible for paying the bill. The larger benefits come in aggregate by proving to utility companies that more

Figure 3.1. Denny Park (Photo courtesy of Enterprise / William Stickney)

housing does not need to result in larger or new power plants. Housing that demands less energy reduces the cost of a solar installation because a smaller system can meet the demand. In Massachusetts, we piloted and then the state adopted a solar-ready pathway, allowing developers to build green affordable housing that was ready to accept a solar array when it became affordable. That is, properties that were designed to be more energy efficient from the start resulted in properties that would require smaller, more affordable systems that could be leased or purchased as the market allowed.

Also worthy of discussion are the environmental benefits in building operations. When a building is designed to deliver proper comfort, adequate ventilation, daylight, and access to natural areas, there is less need to rely on energy inputs and products to maintain the building or overcome the deficiencies. There is less need for air conditioning in the summer or the heater in the winter. When there is proper ventilation there is less need to run a fan or plug in the humidifier. When we design with nature and consider the environment, we can achieve higher levels of performance more easily. Additionally, we are not creating other problems such as moisture buildup or mold. Green housing is more elegant in how it performs. When we use the same building design in every building located anywhere in the country, we are unable to realize these environmental benefits. We will find ourselves fighting our natural environment.

Engineers working with the architects can save costs by properly sizing the mechanical systems for the specific building and location. Too often heating, ventilation, and air-conditioning systems are recommended based on the square footage of the house rather than architects and engineers maximizing the building's design and orientation to lower the overall need for heating and cooling to downsize the mechanical systems. Doing so can save not just money but also space. In some circumstances it can result in space for another small room or closet, thereby improving the overall comfort and livability for residents.

There was a wide range of costs for both electricity and water. The cost for electricity ranged from a low of $0.03 to a high of $0.24 per kilowatt-hour, whereas water ranged in price from under $3 to over $16 per thousand cubic feet. Clearly this has a significant impact on the life cycle value of the savings. It was clear that the projects with the lowest return on investment and the longest payback periods were the ones with the lowest costs for utilities. Utility costs are subject to high levels of volatility and potential for price shock. Although energy is usually in the headlines, water rates are perhaps more likely to be varied and changeable. Many communities have substantial deferred maintenance issues on their infrastructure and will need to make major investments in the coming years. These will result in very sharp rate increases in many locations, often at least doubling the cost of water. Energy and water conservation reduces the long-term costs and the properties' exposure to sudden price changes. The findings from the 2012 study largely mirror those from the 2009 study in that lifetime financial benefits from implementing the Criteria largely cover the cost of their implementation.

Green affordable housing can ameliorate some of these concerns. For example, if you are monitoring your water and energy usage in real time, you will quickly detect any problems. This will prevent such things as long-standing leaks that can lead to water damage and structural issues. It also prevents unexpected spikes in utility bills forcing you to forgo spending on other critical needs or to withdraw from reserves. Because green affordable housing is now requiring more active and passive measures to adapt to the impacts of climate change, owners can have greater confidence that in facing an electrical outage or massive weather event, the property will be prepared to handle it. Climate adaptation was not a part of the initial Green Communities Criteria, but in 2015 it became a necessary feature. It requires that the design team carry out a vulnerabilities assessment and implement building elements designed to enable the project to adapt to, and mitigate, climate impacts given the project location, building and construction type, and resident population.

People Are Healthier

Many additional benefits gained from integrating the Criteria into housing do not have direct measurable financial impacts to developers or owners, but these also typically pay for the improvements. These include better occupant health and well-being through reduced exposure to environmental pollutants, improved connectivity to services and walkable neighborhoods, and good daylighting. The benefits extend beyond the occupants to the neighboring community by supporting local community services and providing activation of the neighborhood streets, improving water quality, and reducing the impact of rainwater runoff on neighboring sewer systems and water courses.

In general, developers found ways to integrate green measures into their affordable housing designs and were able to implement them in cost-effective ways. Developers were able to meet a number of Criteria for no additional cost, and even where a cost premium did arise, the overall impact on the budget was small. The median cost to meet all mandatory site selection location and site improvements Criteria was $0. This is not surprising because most of the projects were built in areas where local and state building codes already mandate that projects meet many of the measures included in the Criteria. The median cost to meet the water conservation Criteria was $83 per unit, and the median interval before payback was less than two years. Costs to meet the mandatory energy conservation Criteria ($1,056) accounted for the majority of the premium associated with meeting the Criteria. There was a small cost premium related to materials selection, a median of only $165 per unit. The cost to meet the Healthy Living Environment section of the Criteria accounted for the second-highest cost premium, with a median cost of $680 per unit. However, 14 of the 52 projects reported no cost premium to meet these

Criteria, suggesting that some projects were able to incorporate these particular green measures within their existing budgets.

Many affordable projects know the importance of healthy living environments. Sanctuary Place in Chicago, mentioned in Chapter 2 as a property that embodied the Green Communities Criteria before they had been created, was completed in 2004 to provide housing and counseling services for women on a campus that provides the opportunity for women to reunite with their families after having been separated as a result of being unhoused, incarcerated, or addicted to drugs. Gladys Jordan, the president of the Interfaith Housing Development Corporation, was well attuned to the problem of indoor chemical exposure because the women who would be living in Sanctuary Place often had compromised immune systems.

What is so remarkable about Gladys is that she understood the importance of providing healthy housing because she understood the needs of the women she was serving. She wasn't developing housing *for* them, she was developing their housing *with* them. The women had extremely low incomes, many had been unhoused, many had HIV/AIDS, and many had been incarcerated. The codesigned vision of Gladys and the women for meeting their housing needs was not simply to shelter them but also to provide them a home that would be supportive of repurposing their lives, reuniting them with their families, and taking care of their health. To do this, Gladys hired architect Kevin Pierce, who had experience working with green materials and creating healthy indoor living environments. He helped bring to life the type of housing that would put the women calling Sanctuary Place home at the center of its design.

Kevin recognized the financial constraints the project was under and provided designs that focused on the physical and mental health benefits. The shared hallway has a secured closet near the door to each apartment to save space in the unit's interior. Each apartment is less than 400 square feet to accommodate more residents and so that parking spaces did not have to be provided on site, as residents would not have cars. Most importantly, each apartment had large, energy-efficient windows to let in light but keep out the cold air of Chicago winters. Natural daylight is important for everyone, and perhaps more so for people who have been incarcerated and for those dealing with mental health issues. Using natural daylight is a defining feature of green building and is a method for not only conserving electricity but also improving the indoor living environment.

Gladys had secured funds for energy-efficient materials and renewable energy systems from the Illinois Department of Commerce and Community Affairs, the Illinois Clean Energy Community Foundation, and the Chicago Department of the Environment. The materials that went into the project had been carefully chosen to be beneficial to the environment and occupants' health. Minimal energy and water resources were needed to operate the homes. The stormwater that ran off from the roof or the sidewalks was being managed on site to recharge the aquifer and not overburden the obsolete infrastructure of the community. Today, in some markets, such a design feature might qualify for stormwater credits that would bring additional financial incentives to the development.

Sanctuary Place was one of the early projects to receive accolades for being green, including winning then mayor Daley's 2004 GreenWorks Award for "outstanding residential project." Many green affordable housing developments receive awards, not always for being green but for being symbols of great design and stewards of public and private resources. They outperform gray housing for no additional or only minimal upfront costs that are often paid back in a few years from operating cost savings.

So, although as a country we still need to rid exposure to known poisons like lead, we have only begun to address other known classes of toxic chemicals that continue to be used to manufacture products we use in our homes. The promise of green building to deliver health benefits requires those involved in designing, manufacturing, and purchasing materials to consider the life cycle of a product's health effects, from the extraction of its ingredients through the manufacturing process to its use in our homes, and how it can be reused or recycled. In the state of Washington, a worker in his mid-30s developed occupational asthma from installing spray foam insulation in residential attics and was forced to leave his job, a common occurrence. Spray foam insulation contains isocyanates, chemicals that can cause asthma and are toxic to the respiratory system.[2]

Fortunately there are many proof points about the health benefits of green building. One of the most compelling studies that has been replicated is based on an effort by the Seattle Housing Authority and the public health departments of Seattle and King County along with Neighborhood House, the University of Washington, and Enterprise Community Partners. Here, 60 "Breathe Easy Homes" were built when High Point was redeveloped. All of the homes within High Point were built with features to improve indoor air quality, which, as mentioned in chapter 2, informed the final creation of the Green Communities Criteria. The Breathe Easy Homes intentionally went further to better serve a population of known asthmatics.

The Breathe Easy Homes were constructed in ways that help further decrease the risk factors that cause asthma. Features include positive pressure house ventilation systems with air filtering, tempered fresh air supply, and heat recovery to improve indoor air quality; linoleum flooring in living areas and bedrooms, recycled-content vinyl flooring in bathrooms and kitchens, and low-pile carpeting on stairs and in hallways to reduce allergens; low/no off-gas trim and millwork to reduce urea formaldehyde bonding agents; low/no off-gas or volatile organic compound cabinet construction; HEPA-filter vacuums to remove allergens; walk-off doormats to reduce dirt in the homes; construction sequences that include extra dry-out time to minimize mold growth; weather protection of on-site materials; ductwork protection to minimize dust; extra clean-ing cycles with low toxicity / nontoxic and nonallergenic cleaners; and flushouts to allow the evaporation and off-gassing of materials.

Residents of these homes participated in a scientific study that docu-mented their health outcomes. The study found that asthmatic children in Breathe Easy Homes have 63 percent more symptom-free days than in their previous homes and showed dramatic improvements in lung func-tioning. Improved health resulted in a 66 percent reduction in the need for urgent medical care. As asthma symptoms and triggers declined, families' quality of life improved, with fewer sleepless nights, less lost work and school time, and lower medical expenses.

The Green Communities Criteria went beyond requiring measures that benefited only the health of the resident, but those types of Crite-ria that had wide-reaching benefits were the hardest to defend because the additional cost or effort was seen to affect the "wrong pocket." That is, the person expending the effort or incurring cost was not receiving the reward. The benefits that directly improve the health of residents are seen as easier to justify because it is possible to apply for grant funds to pay for those items and to form partnerships with health institutions that might pay for additional costs if they resulted in fewer household asthma triggers and thus fewer visits to the emergency room.

The Planet Is Better Off

Meeting the Criteria will significantly reduce energy consumption and thereby reduce operating costs and the utility expenses for residents and operators. However, I also wanted to demonstrate that green afford-able housing could be part of the solution toward driving down carbon emissions in the United States. Investing in additional carbon emissions reductions from affordable housing can bring needed resources to com-munities who otherwise might not be able to benefit from renewable energy systems or deep energy efficiency. Because the Criteria established a uniform baseline of what all Green Communities developments had to meet regarding energy consumption and associated carbon emissions for specific properties, we could then calculate what measures could be taken to provide additional reductions.[3] We knew that developers and owners who were meeting the Green Communities Criteria were well positioned to further reduce carbon emissions in their developments, but they did not have the budget to do so. Many of the developments were built to be ready for photovoltaic systems, and they installed them as soon as it became cost-effective to do so. However, for many, solar energy was simply out of reach. Myself and Stockton Williams at Enterprise acquired a protocol for an offset fund. We then hired Performance Systems Development to help us adapt the protocol to the affordable housing sector. We did not want to sell offsets to anyone who was not totally committed to reducing their footprint as much as they could. So we decided to sell offsets only to entities in the green real estate business.

Through the Green Communities Offset Fund, we proved that affordable housing can reduce its carbon emissions with the resources to do so such that the developments can be verified and monitored over time, even to the world's most stringent standards, which at the time was the Gold Standard.[4] This was a major milestone. Proving that affordable housing developments could reduce carbon emissions meant that any federal or state commitment to climate action must then include affordable housing. Instead, what often happens is that affordable housing is viewed as too complex or fragile, and it is exempted from policies and regulations that otherwise might improve this segment of the built environment. Over time, the housing stock is seen as inferior or a burden on society rather than as part of the future of the country and part of the solution to addressing wicked problems like climate change. The Green Communities Offset Fund demonstrated that with the right tools and resources, we could provide more affordable housing and lower our carbon emissions, essentially planting two trees with one seed. Affordable housing is an essential thread in the fabric of this nation and can contribute to meeting our collective vision for a vibrant economy, and a flourishing planet.

Through the Green Communities Offset Fund,[5] it became possible for owners of affordable housing to benefit from a global carbon market in ways that have the potential to bring new financing vehicles to a sector in desperate need of them. Affordable housing is an integral part of the solution to reducing carbon emissions and must be considered as such. If not, it will be relegated to the sidelines during our new era of anthropogenic climate change—and the biggest threat to our humanity. We must stop pitting affordable housing against the environment or against the economy and realize that it is beneficial to both.

The first large offset we sold was to the US Green Building Council (USGBC) to cover emissions from an expo they hosted at the 2008 GreenBuild Conference in Boston. It allowed us to purchase 330 tons of additional carbon emissions reductions from a development in Albuquerque, providing supportive housing for formerly homeless individuals.[6] That development achieved the additional reductions by installing solar panels to provide electricity to the property. Considering that the average US household produces 7.5 tons of carbon dioxide per year, it is the rough equivalent of 45 households not producing any carbon dioxide for a year.

To make the deal work we bought five years of emissions reductions from the development with proceeds from the offset sold to the USGBC and raised philanthropic funding to monitor and validate that the reductions did take place. Because we were committed to setting high standards using proven and existing methodologies that could not be called greenwashing, we followed the Gold Standard protocols. This was an expensive transaction for us at Enterprise, but it was worth the expense to demonstrate that with the resources, the developers and owners of affordable housing are capable of reducing carbon emissions and that projects in the United States can benefit from this financial vehicle. Enterprise is no longer operating the Green Communities Offset Fund, but it could be resurrected if the transactions could be lowered for the measurement and verification efforts required to prove that additional reductions occurred. However, offsets are a dangerous panacea. If we engage in their use, we must ensure they do not allow pollution to persist in some communities while paying for environmental benefits to accrue to other communities at their expense.

Another area where green housing delivers measurable environmental benefits is in specifying and installing more durable materials that do not just look good but also perform for the people using the building. This means less or no carpeting, and instead using area rugs, where they make sense. This means surfaces that are easier to clean. This means using plants that will thrive and will require little to no watering. More durable materials mean less waste, fewer materials headed to the landfill, and fewer new materials needed to replace those that did not last or did not wear well.

Where housing is located will have a direct impact on the transportation options available to the residents. Choosing smart locations results in shorter commute times and lower costs for residents and lower carbon emissions from transportation. These Criteria are about decreasing the development's impact on the natural ecosystem and ensuring that any new housing developments are located in existing communities and improving connections to amenities through sidewalks and proximity to mobility options. The Green Communities Criteria include location and neighborhood fabric Criteria because of the many ways the location of housing can reduce other household expenses for occupants and because these Criteria were nonnegotiable for the environmental community. Conservation groups would not recognize housing development as being green if it was not compact, with connections to existing development, services, and infrastructure. This can lead to heated debate in some places.

For example, one developme...
mile of a major bus stop and a h...
idly travel to downtown and to job ...
to ensure extra air filtration systems to ...
entered the building from the highway. Th...
our nation needs a complete transition to gr...
and our buildings are providing environmental and ...

Keeping existing trees and vegetation can provide sig...
mental benefits. These trees are often sequestering carbon dio...
tecting the site from erosion. The High Point development in S...
to put signage on all the trees where the redevelopment was occurr...
ensure that they were not accidentally bulldozed. The signs included t...
monetary value of each tree to drive home the importance of keeping them
alive. I have been to many green developments where the trees have their
own story. They often become the focal point of the development, as is the
case for the Phoenix House in Atlanta, where the main courtyard around
which the homes sit includes a giant tree that provides shade and reduces
ambient noise.

Where Are the Criteria Now?
From 2005, when we at Enterprise released the Green Communities
Criteria, to 2020, there have been 127,000 green homes created or renovated; $3.9 billion invested in the development and preservation of green
healthy homes, hundreds of housing organizations supported across the
country, and policies transformed, including qualified allocation plans
for the Low-Income Housing Tax Credit and numerous federal housing
programs.

The Criteria were never intended to live in stone. The notion was
to offer a new way of thinking about affordable housing that would
create a baseline from which we could move an industry forward. The
Criteria were improved over time as we learned what was possible and

. delivered a
›latform for
tal benefits.
ıventionally
ferent fields
›pers to con-

in 2005, the
5, and 2020.
·ovided some
ling the scope
ıte more sus-
ı was to have
v or existing,
ıily detached.
was remedied

it in California was located within a quarter
eeway entrance so the buses could rap-
centers. However, the developer had
mitigate the toxic emissions that
s is another example of why
en so that our vehicles
health benefits.
nificant environ-
xide and pro-
attle had
ng to

in 2008 was the need to better consider the needs of existing housing. This was a time when we experienced significant foreclosures, and new construction was almost at a standstill. This was exacerbated by the fact that the main driver for financing affordable multifamily housing is the Low Income Housing Tax Credit, which requires entities with tax liabilities to invest equity. Major financial institutions involved in the residential mortgage business did not have the need to offset tax liability and therefore were not interested in paying full price or any price for tax credits. Many developments were halted in their tracks. With so many foreclosed properties, the affordable housing community was working to respond quickly to stabilize neighborhoods. The silver lining was that it allowed the Green Communities Criteria to be integrated into the housing rehabilitation process. The 2008 Criteria reflected this need to pay more attention to the specific needs of existing housing. On the energy side this meant allowing for energy ratings to be established and thresholds for how to exceed those in the process of retrofitting the structure. The 2008 update also included a careful consideration of what was possible regarding mechanical ventilation, which is much more difficult to provide in existing houses than it is to design into new ones, but it is no less important and perhaps even more of a consideration when the building envelope is being tightened.

The 2011 update to the Criteria was done with an eye toward clearer instruction and easier pathways for the different types of development. The 2011 Criteria responded to developers who were delivering housing in communities that fell outside of a neat definition as being urban, suburban, or rural. It was our response to the unevenness of how jurisdictions apply planning and zoning authority and how they allocate capital improvement dollars for infrastructure. It was in response to the growing interest from Indigenous communities and a new effort that emerged from a Rose Fellow's relationship with the Santo Domingo Pueblo.

Enterprise updated the Criteria again in 2015 (after I had departed to join the JPB Foundation) to reflect the growing interest in how to prepare for the impacts of climate change. The update required that community health and relevant climate resilience measures be included throughout the integrated design process. Furthermore, it closed the loop between the energy and water savings to influence how the properties were operated to maximize those benefits. These changes reflected what we at Enterprise were learning from the technical assistance providers and development teams regarding areas that could be strengthened. As more developments met the Criteria it was easier to collect input and update the Criteria to ensure that they remained relevant and provided sufficient guidance as to what was expected and how to meet each criterion, based on the experience of completed projects.

Enterprise again updated the Criteria in 2020 under the leadership of Krista Eggers. The purpose of these updates is to keep the housing field moving forward, always integrating the best of what we know to

benefit the lives of low-income persons while keeping housing affordable to operate and maintain and adapting to our changing planet and
the needs of our communities. The purpose is not to make it more
difficult to meet the Criteria. Through the updates, the Criteria can
evolve to reflect what the field is learning and able to achieve with the
entire house, property, and community in mind to leverage benefits for
people and the planet.

The Influence of the Green Communities Criteria

The Criteria remain *the only standard* designed specifically for affordable
housing of all types, from single to multifamily and from rehabilitation
to new construction (see Figure 3.2). Based on an evaluation of the
Enterprise Green CommunitiesSM Initiative by Ross Strategic, released
in 2018, there is no refuting that the Green Communities Criteria played
a critical role in greening the affordable housing sector.

The biggest triumph for advancing green affordable housing came
about when states (e.g., Minnesota and New Mexico) started requiring
that the Criteria be met to gain eligibility for the state's allocation of Low

National & Regional Green Building Standards

Standard	Scale	Exclusive to Affordable Housing
Enterprise Green Communities	National	Yes
U.S. Green Building Council LEED	National	No
ENERGY STAR	National	No
National Green Building Standard (NGBS)	National	No
Living Building Challenge	National	No
Net Zero (NZEB)	National	No
Passive House	National	No
WELL Building Standard	National	No
Green Globes	National	No
EarthCraft	Regional	No
Earth Advantage	Regional	No
GreenPoint Rated	Regional	No

Figure 3.2. National and Regional Green Building Standards (Source: Ross
Strategic's Enterprise Green CommunitiesSM evaluation)

Income Housing Tax Credits. Enterprise was one of the early thought leaders behind the idea for the tax credit. As such, it is no surprise that the organization set out from the beginning to use the state Qualified Allocation Plans for how developers could access tax credits as a mechanism for incentivizing developers to integrate the Criteria into their developments. Although it took working with the Housing Finance Agency staff and the development community state by state, it was one of the most impactful strategies undertaken by Enterprise and continues to be the best vehicle for incentivizing, or in some states requiring, the Criteria to be met. Because they constitute a comprehensive platform that has been applied in every type of new or existing housing development, local, state, and federal government agencies can use the Criteria to further specific agendas around health, energy, or design. However, relying on this one program is not enough. Between 1987 and 2018, according to the Low-Income Housing Tax Credit database from the US Department of Housing and Urban Development (HUD), only 3.23 million units were placed in service as a result of these tax credits, a fraction of the housing needed to keep up with demand. Without an expansion of this program alongside additional financial tools, we will continue to lose ground on providing the amount of housing needed at differing price points.

The federal government used the Criteria as the standard for accessing the 2009 American Recovery and Reinvestment Act dollars, for neighborhood stabilization dollars during the mortgage foreclosure crisis, and for numerous HUD programs, including HOPE VI. Having a cohesive set of Criteria allowed a simple way to advocate for compliance with the Criteria in new funding streams and housing programs. Developers are typically keen to win new development projects and will agree to most requirements if they are understandable and achievable within the given cost constraints. For example, the enormous new Beltline initiative in Atlanta opened up opportunities to build new housing and to rehabilitate existing housing within that designated area. This allowed the City to stipulate that any developer seeking to build there must meet the Green Communities Criteria. For example, the previously mentioned Phoenix House had to meet the Criteria when it was rehabilitated in 2017 because it was located in the designated Beltline area.

The Criteria that people thought were impractical or too costly to implement in 2004 are now commonly practiced in many geographic locations. The biggest triumph of all is that the Criteria have become a platform from which to address the challenges facing the affordable housing sector and to take advantage of new opportunities. The Criteria also level the playing field by establishing a threshold for what is considered to be quality housing. In the beginning, even in places like Austin or Portland, where the development community may have been accustomed to implementing measures around water conservation or energy efficiency, no developers were thinking about all aspects of the Criteria, from integrated design to healthy materials to operations and maintenance.

The Criteria established that green methods and materials had to be considered throughout the entire development, from choosing where it would be located to educating residents on ways they could support the green goals through their use of energy, water, and choice of cleaning products. Aeon, a nonprofit community development corporation in Minneapolis, was an early adopter of the Green Communities Criteria and participated in the Minnesota Green Communities Initiative to ensure that the Criteria could be applied in that region's climate and construction culture. As a result, Minnesota has cultivated a thriving green affordable housing industry. This early commitment to green building has also created strong movement toward a just energy transition in the state to prepare communities for the impacts of climate change while mitigating future impacts by reducing carbon emissions.

In one particular Aeon property, the Wellstone, the residents were informed about the green goals for the property but were not particularly engaged. They had busy lives and families. When they learned that the building was designed to conserve water, however, it resonated with many of the residents from Somalia, who had experienced severe droughts. Through conversations on the topic, residents who thought that by hand washing their dishes and clothes they were saving water realized that they were using more water than residents in other buildings. They learned that using the dishwasher and laundry facilities would actually save water, so they switched from hand washing, and they tracked their lowered water usage.

The Criteria as Living Platform

The Criteria engage everyone from the design team to the construction workers and their subcontractors and from the residents to the maintenance staff. Each new development or rehabilitation of existing structures is an opportunity to strengthen neighborhoods and to connect the community members to one another, both those who are living in it and those who are working in it. Although the Criteria do have a checklist of the items that must be met to be considered a Green Communities development, the Criteria themselves are intended to be a living platform to rethink how we are locating, designing, and delivering housing. By engaging with everyone involved in the process from the initial discussions about which parcel of land to purchase, it is incumbent on us to listen to the community. In the case of the Aeon property, the community already wanted to use less water because they knew that water is our one nonrenewable resource, unlike energy. Although the property was designed to use less water, because it was meeting the Criteria, perhaps it could have done more, knowing that it was a priority for the residents. At the very least, communicating with the residents who did move into the green property allowed them to understand what they could do to further the green goals and conserve water. This became a point of pride and healthy competition among the residents.

The experiences of the residents at the Aeon property are similar to those of others across the country. Maybe it is fundamental to our human existence that we want to be part of something bigger than ourselves. We want our actions to do more than just benefit ourselves. The Green Communities Criteria are a signal to the world that we understand the impact that development has on the planet and people, and although we must urgently increase the supply of housing accessible to people with low incomes, we can do it in a way that is better and more responsible and more fulfilling for all involved. The robustness of the Criteria requires those involved to see the development as a whole, whether it is new construction or a rehabilitation project. Consider this: If you want to reach a certain level of energy efficiency for the building but do nothing to address water conservation, then every time hot water is used more energy will be needed to heat more water. However, with water conservation measures in place, less water will be used, so when hot water is needed, less of it will need to be heated. Depending on the energy source, in aggregate, significant carbon emissions can be avoided. If developers want to deliver a healthy property and install proper mechanical ventilation to keep the air circulating, with fresh filtered air, but we allow unhealthy furnishings to be installed, then we are compromising the health of the building and the residents. The costs of the ventilation or for the energy efficiency in either scenario cannot then be justified because neither ventilation nor energy efficiency will be realized to its maximum capacity as it otherwise would be if a more robust approach had been taken, as the Criteria require.

By establishing the Criteria as a threshold for moving us forward on all aspects of green development, we have a platform for continual improvement. I do not think any of us would trade in today's mobile phone for the first one we purchased. Moving the world to mobile technology has allowed us to imagine and deliver so much, from mobile health services to less constrained transportation alternatives. The Green Communities

Criteria brought together environmentalists with affordable housing advocates because of the Criteria that address ecologically sensitive site improvements and smart locations. The Criteria brought together the booming energy efficiency and renewable energy sectors to figure out the ways to serve the affordable housing sector. The Criteria brought together the health care industry and the housing professionals to find new financing vehicles for investing in green affordable housing and demonstrating ways to divert what would have been operating expenses for energy, water, and maintenance into resident services because the building finishes are more durable and the utility bills are lower. Focusing on only one aspect of the Criteria diminishes the overall benefits and can have unintended negative consequences.

The Green Communities Criteria have always been built on proven methods and materials to deliver more housing affordable to more people. With each update to the Criteria advances have been made to clarify methods or to be more accessible to developers and owners working in rural areas or with existing properties on tight parcels of land. Additionally, optional items exist to allow those who can to stretch and go farther to deliver health, economic, and environmental benefits while producing as much housing as possible, making it accessible to those who need it the most.

To make change you need to have a legitimate platform from which to seek the change. You need to offer the framework and road map for others to join you; there must be the early adopters and trusted advisers to validate that change is beneficial; and then, for everyone else, if incentives are insufficient, it will require policy and regulation. With at least 127,000 Green Communities homes, we know that the Criteria can be met. The exercise now will be to keep the Criteria alive and reflective of what we know to be best practice so that we can continue delivering measurable benefits. The Green Communities Criteria have allowed us to define green in a robust way, turning it into a platform for delivering multiple benefits. When we start pulling the Criteria apart, we water down our collective understanding of what it means to be green. A building can be energy efficient without being green. Think of a net-zero energy building located on an island in the Caribbean that requires taking a private jet to get there. There is nothing green about that scenario.

The Green Communities Criteria are a way of thinking comprehensively about what the design, construction, rehabilitation, operations, and maintenance of affordable housing can do to improve the lives of everyone involved and to benefit our planet. This is the big shift. I witnessed it firsthand when we first rolled out the Criteria. The initial hurdle was considering all aspects of the Criteria. It was not insurmountable, but doing so was not perfunctory. With intentionality to consider location, sensitive site improvements, indoor air quality, energy, water, operations, and maintenance, you begin to think differently. You involve more people in the decision making, you consider different perspectives, and you deliver a better outcome. It is no coincidence that the properties that typically receive best-in-class recognition are green buildings. The properties that make it onto the cover of the community development corporation's annual report or onto the conference brochure are green developments or people living in green buildings.

Because we now have a more robust way of thinking about affordable housing development, by using a new vocabulary, starting with "integrative design," we can advance and not become mired in delivering cookie-cutter housing that could be placed anywhere but not function well for anyone. Although I led the creation of the Criteria with climate change in mind, I know I am on the record saying, even in 2011, that "climate change seems far off." In 2011 it was not far off for communities in Alaska, in Louisiana, around the Great Lakes region, and in many cities and countries around the world. Less than 10 years ago we were thinking about climate change mostly in terms of how to mitigate it by lowering our demand for energy and products that relied on fossil fuels. Today, not only do we need to ramp up our climate mitigation responsibilities; we also have to quickly and responsibly deliver housing that is adaptive to our human-induced changed climate. Because the Criteria are applied to all housing, new and existing, they create the right platform from which to ensure that the communities most impacted by our vulnerable systems are not left even further behind because the houses where they live are not equipped to handle intense weather events, power outages, and extreme heat and cold.

The 2020 Green Communities Criteria include the following rationale for considering the impacts of climate change in how a housing development is designed, constructed, and rehabilitated: "We live in a world that has already warmed 1 degree Celsius (1.8 degrees Fahrenheit) over preindustrial levels thanks to climate change. Science finds that higher temperatures are causing a range of worsening impacts, and we're already seeing damages. Low-income communities are on the front lines of climate change and often have the least access to resources needed to recover from a disaster. With the increasing frequency of storms, floods, and other extreme weather events, the costs associated with not investing in resilience are rising rapidly. Investing in resilience before disaster strikes is one of the most cost-effective ways to protect residents and property while strengthening their ability to weather the increasingly severe storms ahead."[7]

Addressing climate change in housing is not just about stilts and mechanicals on higher floors. There is a much more human element to it as well that involves fostering community cohesion by ensuring that our housing offers spaces to gather to interact with one another. It involves strengthening connections to the existing fabric of the community through smart locations and enabling the culture of the residents to flourish. It also means that we have to continue to be diligent about the type of materials and products we use in our housing. We have to do our part to stop demanding products that continue the proliferation of the toxic chemical companies that destroy everything around them while exposing us to poisons. Not only do these products off-gas and enter into our bodies without our permission; if there is a disaster, whether flood or fire or even a period of time without electricity, our exposure is amplified. The products containing the chemicals degrade and expose us to these toxic elements, and if we are trapped inside with them, we are more exposed to their off-gassing.[8]

In our construction and rehabilitation of affordable housing, we will have to consider the impacts of climate change. There is no excuse anymore for those in the real estate sector not to consider both how they can mitigate climate change by not increase carbon emissions and how they can use the development to better prepare and adapt to the changing climate. Fortunately, affordable housing developers are more attuned to it because of Green Communities.

The Criteria have evolved to consider how the property would respond to a weather-related event and to consider climate change at the integrated design phase. The platform is such that it can evolve to help an entire industry continue to think differently about how it delivers affordable housing.

Conclusion: The Heart of the Matter

The Green Communities Initiative puts affordable housing on the green building map. It made the case that green mattered to this sector as much as it did to the commercial building sector. Committing to green affordable housing goes beyond the immediate benefits to residents and the owners and operators of the housing. For everyone in the industry going green signals our acknowledgment that even in this "do-good" part of the work, we know that we are part of the problem. We are demanding products and resources that without much thought can lead to deforestation, toxic waste, and climate change. The Green Communities Criteria were designed to provide a low- to no-cost way to do better for the planet without sacrificing the number of affordable homes that could be constructed or rehabilitated.

We must think about our homes comprehensively, as if they are living organisms that are interconnected, because they are. If you have a car running in the garage without any exhaust fans or barriers between that space and the indoor living space, carbon monoxide will pollute the indoor air and put at risk anyone inside. If there is radon in the land below the building, with no cap or sealant between that space and the home, then people living there will be exposed. If we tighten and seal a home to make it more energy efficient but neglect to ventilate it, then people will get sick. If the design team has no experience living in the community or connection to its culture and people, then they will design a building that is at odds with the needs of the people who will be living there. If no one has engaged with the residents or maintenance staff to share how the building is designed to perform, then they will not be able to deliver on its expected outcomes. The Criteria are simply a platform to elevate the fact that a house can be more than four walls and a roof if there is an intentional signal set in the very beginning to think comprehensively about delivering health, economic, and environmental benefits for all. This intentional signal must be strong enough to be known and understood by all involved.

CHAPTER 4

The Challenges to Greening Affordable Housing for All

All of this requires us to recognize a power greater than ourselves and a life longer than the one we will live. We must transform from a disposable, short-sighted reality of the individual to one that values the long-term life cycle of our collective humanity. Even the best of us are tangled in an unjust system. To survive we will have to find our way to a shared liberation.

—Collette Pichon Battle

Even with the success of the Green Communities program, we have a long way to go before we can make the claim that all housing is green. Given the immense need to preserve and produce more housing over the next 10 years and the simultaneous need to limit carbon emissions to at least 450 parts per million to maintain global warming below 2 degrees Celsius, we must take action on both fronts. The health of people and our planet depends on it. Not doing so will impact all of us because we are connected, but there are challenges to doing so that require our attention. In my experience, there are four major challenges hindering our success to advancing climate action by addressing our housing crisis using green methods and materials: (1) inexperienced and unavailable workforce, (2) lack of transparency about materials ingredients, (3) uncertain and scattered financing, and (4) weak national commitment.

Inexperienced and Unavailable Workforce

The 2020 US Energy and Employment Report tracks changes in energy and energy-related employment. Among construction employers in the energy efficiency field, one of the largest surveyed sectors with over 1.3 million workers, 91 percent of employers reported that it was somewhat difficult or very difficult to hire new employees.[1] Employers cited lack of experience, training, or technical skills as the top reason for difficulty hiring. The challenge of a constrained labor market is that projects can be delayed, and when a project team does not have the green building experience it needs, then it will default to non-green practices, or it might charge more to cover the risk of the unknown, and it may incorrectly install systems leading to lower-than-expected building performance.

Furthermore, smaller contracting firms may not have access to lines of credit so that they can complete projects and then get paid. This may also mean they cannot afford to train potential employees on job sites because cash flow may be too tight. We lived through the constrained contractor market during the American Recovery and Reinvestment Act of 2009. We witnessed large firms' ability to quickly ramp up and manage the uncertainty of project approvals and financing. A large firm is more likely than a small one to be approved for a line of credit from a bank and to have a reserve fund that can cover up-front costs until it is replenished once contracts are completed. Furthermore, larger firms may not be spending profits locally as they may be based in a different city or state from where the projects exist. These same firms are also unlikely to be minority- or women-owned, meaning wealth will accrue to populations not representative of those living in the developments where the firms are working. One example of an effort aimed at helping small, diverse contractors grow and sustain their economic viability is Elevate Energy's diverse contractor accelerator program. Elevate Energy implements programs nationwide that ensure the benefits of clean and efficient energy and water use reach those who them most. The contractor accelerator program identifies diverse contractors, provides technical small business services and back office support, and provides access to clean energy projects and bridge financing. This is a critical set of services to ensure minority- and women-owned businesses flourish and benefit from efforts that support our transition to clean and efficient energy and water use.

We already know that women are underrepresented in the green jobs field. A report by the US Department of Labor's Women's Bureau highlighted the benefits of including women in the green economy. Specifically, it cited that "green jobs provide potentially higher earnings than many 'traditionally' female jobs requiring similar qualification levels; opportunities exist at many skill and educational levels, including opportunities to move up from low-skilled, entry-level positions to highly skilled, higher-paying jobs; knowing that green jobs contribute to a healthier environment can bring job satisfaction; and the green economy offers a higher number of comparably well paid lower- and middle-skilled jobs than the rest of the economy."[2] Fewer than 3 of 10 green jobs are held by women, whereas women make up nearly half of the total labor force.

A just transition to a clean energy future requires new methods for communities and minority- and women-owned businesses to benefit. Meeting our housing needs and reducing carbon emissions create opportunities to provide family-sustaining jobs, not only on the construction site but in the manufacturing facilities—where we can produce healthy products—and through community and large-scale renewable energy installations.

The rapid growth of the renewable energy industry presents an opportunity to build a new energy workforce in the United States with diversity and inclusion at the forefront. Solar jobs present low barriers to entry, with 50 percent of workers attaining no more than a high school diploma, according to the Solar Foundation. However, previous experience is the most important requirement for solar jobs, and in 2019, 83 percent of employers identified difficulties with finding skilled candidates.

Organizations such as the national nonprofit GRID Alternatives ensure that under-recruited communities—including women, people of color, Indigenous tribal members, and reentering citizens—can gain the hands-on experience and skills needed to access careers with long-term potential for advancement and wage growth. They provide training on job sites, which is what a national green jobs corps could provide on a range of green affordable housing sites. Without a robust national training program that prioritizes opportunities for people from the communities where housing is needed most, it will be a challenge to provide green affordable housing with the urgency needed.

Explicit policies and incentives are required to ensure that the expansion of the green economy through increased housing preservation and production includes job training and other pathways for persons of color and women. Internships, apprenticeships, and other ways to pay workers while they learn new skills provide a tremendous opportunity to diversify the green labor force. This includes specific programs to grow a US-based clean manufacturing sector so we have the products and materials needed to preserve and produce green housing.

Lack of Transparency about Materials Ingredients

We spend 90 percent of our time indoors, yet we know very little about the health effects of materials surrounding us. We are hard-pressed to deliver on the health benefits of green housing without having easier access to the ingredients of building materials and products. It is a tremendous failure of public health that chemicals in our building materials are assumed safe unless proven otherwise.

As discussed in earlier chapters, building materials often contain persistent, bioaccumulative, or toxic chemicals. Some are proven (and others are suspected) to be asthmagens, reproductive or developmental toxicants, endocrine disruptors, or carcinogens. Toxic materials endanger residents and can also pose threats to the workers who manufacture, install, and dispose of these products as well as to the communities adjacent to these production facilities and to the broader environment.

In fall 2019, workers became sick after installing carpet in the student apartments at the University of Minnesota. At least one of the workers—Ulysses Eldridge, a subcontractor—did not have health insurance and has not been compensated for the days he could not work as a result of the exposure or for the fact that he is now hypersensitive to chemicals. Tests completed by the City of Minneapolis found that a mixture of chemicals were seeping from the carpet tiles. The chemicals are known toxic volatile organic compounds. Apparently the contract between the workers and the developer had allowed the developer to continue having the carpet tiles installed regardless of the impact on workers' health. To remedy any harm, the developer had the building flushed, meaning it would be ventilated until release of the toxic chemicals dissipated over time. Flushing is a common solution allowed in green buildings where toxic materials can be used if the building is flushed a few days prior to occupancy. Regardless of how well the toxic chemicals have or have not dissipated, it means that the construction workers are exposed. Unfortunately, the underlying problem is the lack of transparency around the ingredients in our building materials, which makes it is impossible to know what to use and what to stay away from.

Recycled content sounds like a good thing. Products with recycled content are typically branded as being good for the environment, which often equates with also being good for us. It might even be the reason why you purchase a product. However, when something toxic is recycled or used as an ingredient in something else, it can perpetuate the toxicity through the production process, sending more harmful chemicals into the air, soil, and water. For example, the use of fly ash in cement might seem like an elegant green building material. How poetic to take one of the most toxic products of the fossil fuel industry, a by-product of coal-fired power plants, and encapsulate it in a building material instead of sending it off into our air and waterways. Nevertheless, when that building is demolished or deconstructed, the concrete is ground up, and those toxic emissions can return to the air, soil, or water, while having already contaminated anyone tasked with grinding it up. Fly ash can permeate vapor barriers and pass into the air of the living space much as formaldehyde does when it is used in insulation. Coal combustion waste contains hazardous substances, such as mercury, arsenic, and cadmium, exposure to which creates great risks. If these hazardous substances are used as recycled or reclaimed content in wallboard, and the wallboard is cut or removed during rehabilitation, the "improvements" can create far more hazardous environments.

This dilemma exists for carpeting too. Many green builders have started using carpet tiles, which allow one tile to be replaced instead of an entire room or hallway of carpeting. However, fly ash is widely used as a filler in carpet tile. The carpet industry has become a repository for the coal industry's waste. Manufacturers have used fly ash to load carpet with so-called recycled content. Some carpet, by weight, is 40 percent fly ash. This has been a marketing advantage: prominent green building certifications have rewarded recycled content regardless of its origin and contaminants. Pollution control devices on power plants transfer mercury, a potent neurotoxin, from air emissions into fly ash. Who wants their child crawling around on carpet with mercury in it? No one. And no one should even have to think about it.

As already described, the climate implications of toxic chemicals are in their production and in the waste stream. The vast expansion of the petrochemical industry is accelerating carbon emissions, and the by-products of that industry are ending up everywhere, including in building products. In 2016, the United Nations Environment Assembly mandated a report, the *Global Chemicals Outlook*, as part of an effort to minimize the harmful impacts from chemicals by 2020. The second edition, *Global Chemicals Outlook II*, was released in 2019 with input from over 400 scientists who concluded that "the world will not meet international commitments to reduce chemical hazards and halt pollution by 2020."[3] These chemicals are impairing human and planetary health because they are causing the death of pollinators and our coral reefs. Greenhouse gas pollution from industries producing the toxic chemicals that manufacturers are, in turn, using in our building materials is warming global surface temperatures.

To stop the production of these materials, we must demand materials without them. This is a massive challenge. Housing developers are ill equipped to research the health of every product that they specify for use in the properties that they rehabilitate or construct. Projects striving to gain certification through the Living Building Challenge (discussed in chapter 2), which has one of the more rigorous standards for healthy materials, must reach out to manufacturers to obtain information about the ingredients of products, which often have very complex supply chains. Engaging in rigorous research on the ingredients of materials adds to the developer's soft costs of the project and could result in delays. Not knowing could result in worker and resident ill health effects and the continued production of materials with toxic ingredients that pollute communities in the process. The challenge is compounded by a weak regulatory environment allowing the use of toxic chemicals in building products, as well as the manufacturing community's lack of disclosure and transparency about the chemicals used in products.

This lack of disclosure and transparency means that people do not understand what they are exposed to and thus do not understand the impacts of the chemicals on human and planetary health. Much like labels on our food, we must have transparency regarding the toxic classes of chemicals that are in our building materials. One pathway to achieving transparency is to require that product contents be disclosed so that anyone specifying products in a housing development can make informed decisions. The Health Product Declaration® Collaborative is a not-for-profit member organization that created an open standard for this type of reporting. Harmonizing how the industry is reporting on its ingredients will provide multiple benefits. Much like we do with our food labels, we can compare products, and we can begin to understand if there is a particular class of toxic chemical that we are using in most products. This allows scientists to explore nonregrettable substitutions that lead to the elimination of that class of chemical. Waiting to see how many people get sick from products and then proving that the sickness is caused by those particular products is not the way to deliver healthy housing. The onus must be on the manufacturer to disclose the ingredients and to prove to us that their products are safe.

Uncertain and Scattered Financing

The housing system is heavily dependent on a few sources of capital, including debt, tax credits, and federal Housing Trust Fund dollars to allow developers to provide housing at price points that make it affordable to persons earning incomes below the area median. To understand the basics behind affordable housing finance, check out an application called Does It Pencil Out?[4] It is built by the Urban Institute and its Assisted Housing Initiative. Incomes are not keeping pace with the ever increasing cost of living. We need the public sector to help address a challenge of this magnitude. Without a holistic financing system coordinated from the federal to the local level in place to support the scale of affordable housing rehabilitation and construction needed, efforts will be hamstrung. Our federal housing finance system is vulnerable. We feel the shock when the government shuts down, when housing budgets are cut, and when investor appetites for low-income housing tax credits change. When financial institutions enabled the subprime mortgage crisis, it exposed the vulnerability of the affordable housing sector due to its almost complete reliance on the Low-Income Housing Tax Credit program to finance the development and rehabilitation of affordable rental housing. All of a sudden, the competition for tax credits was virtually over, and housing subsidies were focused on recovery and preventing foreclosure. Housing construction ground to a halt, which negatively affected workers. Because the banks foreclosed on thousands of homes, there was a spike in demand for affordable rental housing, which did not exist at the scale demanded in locations where it was needed.

From a Green Communities perspective, the situation with low demand for tax credits during the housing crisis was immediately felt on green affordable housing production. Until then, it was a tight market where investors were willing to pay competitive prices for tax credits, which was good for the developer. It resulted in housing finance agencies having a greater level of comfort to require meeting green building standards because the developer would have the equity needed to cover first costs. When the investors lowered the prices they were willing to pay for the tax credits, it immediately put a strain on the development budget because the dollar value for each credit was suddenly lower, and there was less capital available from selling the credits than was expected at the time the development had been conceived and designed. This put pressure on lowering first costs regardless of long-term cost implications on operations and maintenance.

As is true for most disasters, those who have had nothing to do with causing them experience the most disruption. In this case, it was the thousands of people stalled on the waiting list for homes that could no longer be built or rehabilitated because the developers needed financial equity.

Making all of our housing green and affordable to those with low and very low incomes will require a federal housing plan that unlocks the necessary capital to meet the need at the scale required. This will require expanding and improving programs that already exist, such as the Low-Income Housing Tax Credit program, the Housing Trust Fund, and the Community Development Block Grant Entitlement Program and fully funding rental assistance programs that currently serve only one out of every four persons eligible. We must create new financing vehicles, such as those we are seeing through emerging green banks, to finance the full spectrum of green housing needed. Green banks are defined by the Organization for Economic Cooperation and Development as public, quasi-public, or nonprofit entities established specifically to move private investment into domestic low-carbon, climate-resilient infrastructure. For example, the Montgomery County Green Bank in Maryland is able to provide specific debt financing to entities seeking to deliver energy efficient or solar-powered affordable rental housing.

However, when a developer or owner has to go to multiple places to secure the many sources of financing required to deliver the housing, it adds to their costs and ultimately the cost of the project. We need a consolidated effort to bring all the required capital to one place so that we have transparency and accountability around who has access to it, who benefits from lending or investing it, and ultimately how much progress we are making on delivering housing that mitigates carbon emissions. The marketplace has clearly embraced transformational green types of development, but the financing models have not caught up, which is limiting innovation. At a time when we face exacerbating trends regarding housing insecurity, climate change, and stagnant wages and incomes, we should be doing more to meet these challenges with updated tools and ways of working. Continuing to use financial vehicles that favor conventional development is a mismanagement of natural and human capital.

Convincing a Scattered Lending and Investor Community to Prioritize Green Housing

When each developer has to convince every loan officer or syndicator about the merits of green building, then the cost to doing business becomes a challenge. Until the housing finance community fully engages with green building, our best green intentions are at risk of not being realized. Because affordable housing requires subsidies to keep it affordable, there are many financial institutions involved. Developers have to stack together all types of capital, from predevelopment loans to tax credit equity. For example, Figure 4.1 lists all of the capital sources,[5] each with its own requirements and terms, from Trolley Square, an early Green Communities development that offered 40 affordable homes in Cambridge, Massachusetts.

Each capital source has its own risk tolerances—some can incentivize green building and some can be the reason why developers are unable to do so. In some cases, underwriters might calculate it as too risky to move away from elements of a traditional gray development, such as not having a large parking garage even though the development is adjacent to transit, and many of the residents will be seniors or have physical challenges that preclude them from obtaining a driver's license, and some residents will not be able to afford a vehicle. In such a case, the underwriter, rather than focusing on the current population, may instead be viewing it as a risk to resale value if the building ever has to be sold. This is certainly true of mechanical ventilation systems, renewable energy systems, or underground water cisterns. The lender may not know how to value these features or may want to see insurmountable levels of reserves in the event anything breaks down.

Trolley Square Financing

Funders	Type
Cambridge Housing Authority	Rental Subsidies
Cambridge Neighborhood Apartment Housing Services	Deferred Loan
Cambridge Savings Bank	Grant
Charles Bank Homes	Grant
Citizens Bank	Loan
Commonwealth of Massachusetts	Affordable Housing Trust Fund, Deferred Loan
Community Economic Development Assistance Corporation	Pre-development Loan
East Cambridge Saving Bank c/o the John P. Allen Affordable Housing Fund	Grant
Energy Star	Grants
Enterprise Community Partners, Inc.	Green Communities Grant
Home Depot Foundation through Neighborworks America	Grant
Mass Technology Collaborative	Photovoltaic Panels - Grant/Deferred Loan
Massachusetts Development Corporation	Pre-development Loan
Neighborworks America	Grant
New Ecology Inc.	Grant Pre-development
NSTAR	Grants
The Cambridge Affordable Housing Trust	Deferred Loan
The City of Cambridge	Community Development Block Grant (CDBG), HOME, Deferred Loan
The Department of Housing and Community Development	HOME, Low Income Housing Tax Credit (LIHTC), Deferred Loan
US Department of Housing and Urban Development	Rental Subsidies, CDBG, Deferred Loans

Figure 4.1. Trolley Square financing (Re-created from Project Summary: Trolley Square)

First Cost versus Long-Term Costs

With so many stakeholders in the housing industry it is understandable that we want to keep everything as simple as it can be. However, restricting what can be spent up front to cover the costs of green affordable housing is shortsighted. Prevention is always more cost-effective than dealing with the outcomes of inaction. Gray materials that are not durable or healthy for people or the environment will have long-term cost implications. Data are available on what first costs should be. We must let performance drive approvals, not the blunt instrument of evaluating first costs on a per-unit basis. Doing so limits a developer's ability to deliver on long-term savings in the operations and maintenance of the development. It makes it difficult to deliver long-term benefits to the residents and society. We know that wise investments in affordable housing that allow residents access to services, provide community services on site, and improve the health of residents reduce costs to the public sector over time. Evaluating the goals of the development and assessing the costs against the overall benefits will ultimately make our money go farther over the long haul.[6]

Without building performance standards, we limit operators to focusing on paying their bills, not on how much electricity, gas, or water the development is using compared to what it *should* be using. For example, utility bills may be delivered to the property manager and paid without any conversation with the maintenance staff. The bills may be incorrectly pricing the electricity, there may be a water leak, or a photovoltaic system may not be plugged in properly. Yes, this actually happens, and it often takes months for it to be realized and corrected. And all of this leads to a higher-than-needed utility bill and probably higher maintenance costs.

Knowing how much electricity and water each development actually uses, not just the cost of paying for it, can help in creating financial vehicles that better reflect conservation measures. Stewards of Affordable Housing for the Future (SAHF) worked with owners of large portfolios of affordable housing to assess the energy and water usage in buildings.

SAHF hired BrightPower, a provider of energy and water management services for real estate owners, investors, and operators to assess over 280 rental housing properties that underwent energy and water retrofits.[7] It was the first study to examine a large and diverse national data set containing pre- and post-retrofit data for the owner- and tenant-paid energy and water accounts. The study underscored that in the affordable housing sector we were systematically missing a tremendous opportunity by not requiring property managers and building owners to track their energy and water usage along with the associated costs. Simply paying the bill was akin to giving this money unnecessarily to utility companies at the expense of the owner and resident who could otherwise keep it in their pocket if they could reduce energy and water usage. Owners of affordable housing miss an opportunity to save money when they do not manage the demand for energy and water in their properties. If they tapped into technology that allowed them to more easily do so then they could apply those savings to meet other needs of the residents.

SAHF found that one of the reasons for this lack of data is that it can be cumbersome to collect and analyze energy and water utility bills for multifamily buildings. Utility providers have differing and often extensive requirements for allowing access to utility bills, which are the best source of usage data. As a result, while thousands of multifamily properties have undergone energy and water retrofits, actual data on pre- and post-retrofit energy and water consumption are not widely available, especially on a national scale and including unit-level consumption. And for those properties that have not yet been retrofitted, it is impossible to accurately determine where they might find the most savings. They may simply have an undetected water leak or an underperforming system that simply needs a tweak rather than a new part or piece of equipment. Utility programs can better serve affordable housing owners through a one-stop shop approach that helps owners understand energy and water use, provides resources through contractor management and specification support for holistic retrofits, and supports owners with information through the retrofit process. Elevate Energy's work in Illinois exemplifies the success of this approach. Elevate Energy and its core partner New Ecology are mission-based implementers who put the building owners' and residents' interests first. Together they have created the Relay Network,[8] which consists of other mission-based implementers that are working together and innovating in this sector, and collectively retrofitting and reserving hundreds of thousands of affordable homes.

For the SAHF study, BrightPower had to analyze 13,000 separate utility accounts. To bring the data into a manageable format and analyze the results of the retrofit programs, it received utility bill data and energy and water improvement data for each of the programs in spreadsheets. Staff reformatted the spreadsheets and imported the utility data into EnergyScoreCards (ESCs), an online management software, to calculate whole building energy and water consumption for the pre- and post-retrofit years. The energy and water improvement data went into ESCs to aggregate retrofit information into a consistent format for analysis. The program calculated energy and water-use intensity for each property and across the portfolio to determine the level of reductions, normalizing for weather, energy prices, and water prices. They compared actual savings to projections where possible and analyzed the cost-effectiveness of retrofits based on three metrics: simple payback period, savings-to-investment ratio, and cost of saved energy and water, a laborious process.

The level of effort to acquire and analyze these data is extraordinary for many owners who have limited staff and even more limited budgets. What if all utility companies (gas, electricity, and water) had to report their bills in the same format and give them to us electronically, if we requested, and in a format that would allow us to analyze our own data? Managers of large portfolios of buildings need these data as do financial institutions and public sector agencies making the case for green housing programs. To collect these data in real time from properties, we need a national commitment to providing internet service for everyone. COVID-19 exposed that millions of households are without internet service. We must close the digital divide once and for all to ensure equal access to the services provided via the internet that will also have an impact on lowering carbon emissions. Energy consumption data managed via wireless internet connectivity allows utilities or third-party entities to monitor how much electricity is used and when. This can result in energy efficiency improvements that reduce the amount of electricity demanded, particularly at times when utilities must use a generator or peaker plant, increasing emissions if those backup measures depend on fossil fuels. We are starting to see promising innovations taking shape, including the expansion of mesh networks in New York City designed to provide low-cost internet service, which energy management services can then access to track performance in green retrofitted properties.

With affordable housing already difficult enough to finance, developers may not want to take on new ways of designing, developing, or constructing their developments because they do not want to disrupt relationships with lenders and equity providers. Gray housing will continue to be built if it is lower risk, easier, and faster to get approved. It is helpful to have financing for green affordable housing that is designed for green investment where investors want to invest in green and understand the risks and rewards. Financing that recognizes green affordable housing is better and higher performing and therefore is willing to take on perceived risks or provide more favorable terms to the developer.

The Green Communities Initiative helped standardize tools that have enabled lenders to deliver green financing. This includes tools (e.g., a green physical needs assessment) that have been further developed by entities including HUD, Fannie Mae, and Freddie Mac. Fannie used the green physical needs assessment to offer a Green Refinance Plus program. The first affordable housing owner to use that program was LINC housing, for its City Gardens property in Santa Ana, California. Through the refinancing of the property using Fannie Mae's Green Refinance Plus loan, LINC was able to realize $1.5 million for green property improvements that were designed to save LINC and the City Gardens residents' costs on their utility bills. Now we need more examples of these types of programs that focus not only on the energy and water savings but also on the improvements that can lead to better health, without having to access multiple programs.

Lack of a National Commitment

To fully deliver on the promises of green affordable housing we will need a national commitment. We cannot allow any public funding into housing and community development that does not include a commitment to green. We must see our housing challenge in its entirety. Not providing enough affordable housing results in persons being unhoused or unstably housed. Continuing to not take action on climate change means those with precarious and inadequate housing will feel the worst impacts of climate change. As Nan Roman, president and CEO of the National Alliance to End Homelessness, explains, "We know how to end homelessness. Family homelessness has declined every year since 2012. And veteran homelessness went down eight of the past nine years. Now is not the time to abandon the practices that drove those results. Now is the time to get serious about funding them to scale." Those practices that work primarily involve getting people into housing and then figuring out what else they need to remain stably housed.

Affordable housing and homelessness are inextricably linked. We often think of them as different problems. Affordable housing is seen as an income issue—or as a supply issue, where we simply need more physical housing units—or perhaps homelessness is viewed as a problem with the person, who must have made bad choices or who wants to live on the street. These are just not true, however. We have evidence that first providing permanent housing works, rather than first sending people to jail or prescribing services without housing. Providing permanent housing to people first and then engaging with them to support their choice of services needed has proven effective.[9] Even if we address wages in this country, we will still need a commitment to increase the availability of affordable housing.

Conclusion

One of the challenges to meeting this country's housing need is that affordable housing is something that, if you have the means, you do not have to know about it. It is not like transportation, where, whether you are rich or poor or somewhere in the middle, you have to get around and you notice if the roads are full of potholes and if the trains run where they are needed and if they run on time. Transportation infrastructure is seen by all. The population of people without housing is largely invisible to everyone with a home. As human beings, we all share that longing for home and for a sense of belonging, security, comfort, and family; however, if we have a home, and it is in a location of our choice, and we can afford to heat and cool it, then we probably are not reminded every day that at least 11 million men, women, and children do not have a home they can afford. Additionally, 500,000 more do not even have a home at all. In the 2016–2017 school year, according to Child Trends, a nonprofit research organization focused exclusively on improving the lives of children and youth, 1.4 million children were homeless and either staying with other families, in shelters, or in motels, or they were living outside in abandoned buildings, cars, or other places not intended for human habitation. This is unacceptable.

When the COVID-19 pandemic first took over the United States, the plight of those without a place of shelter hit the headlines. In some places, motels and vacant properties were quickly turned into shelters. Many of the people sheltering in their cars or in tents are our essential workers. They had to report to work. The reality that our brothers and sisters had no place to shelter in place became starkly visible during the first month of the pandemic. We must not turn that light away and return them to the shadows of our communities. Everyone must be housed—if for no other reason than that we realize our collective health and well-being depend on it.

CHAPTER 5

A Just Future

When asked if I am pessimistic or optimistic about the future, my answer is always the same: If you look at the science about what is happening on earth and aren't pessimistic, you don't understand data. But if you meet the people who are working to restore this earth and the lives of the poor, and you aren't optimistic, you haven't got a pulse. What I see everywhere in the world are ordinary people willing to confront despair, power, and incalculable odds in order to restore some semblance of grace, justice, and beauty to this world.

—Paul Hawken

WE HAVE AN OPPORTUNITY WITHIN THE NEXT 10 years to provide affordable housing for us all while reducing our carbon emissions, strengthening the fabric of our communities, and improving the overall health and well-being of people and the planet. This entails a just transition to a green future, where everyone can afford a place to call home. Before that is possible, we must acknowledge how we arrived here, such that 11 million people spend more than half of their income on rent and where average global temperatures are dangerously close to exceeding 1.5 degrees, which according to the climate scientists is the threshold at which most natural systems begin to irreversibly break down. What we absolutely must acknowledge is that the impacts of both our housing and our climate crises are disproportionately affecting Indigenous Peoples, Black people, and communities of color. None of those realities happened by chance. Policies, practices, and privilege have legally and illegally created un-level playing fields such that we have allowed a large segment of our society to live housed precariously and in harm's way for extreme weather events and police brutality. Delivering enough units of housing so that everyone is affordably housed will fall short of what is actually needed if we do not consider the larger issues of racial, economic, and climate justice.

When we can point to zip codes and know that children growing up in certain communities will have a shorter life span, which we can determine through the Centers for Disease Control and Prevention, then we are systemically failing to provide socially just conditions where all of us can thrive. When we knowingly subsidize, extract, and burn fossil fuels even though we have been told for decades that doing so will cause climate change, then we are systemically choosing profits in the short term over long-term human and planetary health. When we do not ensure basic levels of income that allow all of us to have an affordable place to call home, then we are weakening the very fabric of our society, on which we all depend.

The value of social justice helps us to recognize the dignity of all people and values every life equally. We have a significant population without permanent housing because of structural flaws in our society, resulting in social injustice. The COVID-19 pandemic has demonstrated that we are inextricably connected to one another in real time. The unhoused person riding public transportation or sitting next to others on the park bench who tested positive for the virus but had no place to recover and quarantine was likely spreading the virus to everyone else sharing those public spaces. However, we should not operate out of fear but rather compassion for one another and recognize that every human being has the right to a home they can afford. To make that a reality we have to build and rehabilitate tens of millions of homes and vacant properties.

Our society needs dignified housing. Essentially this means housing that has access to daylight and that was designed intentionally for the people who call it home, housing that is placed on the land in ways that connect it to the cultural context of the community and that consider its ability to provide protection during intense weather events and pandemics. Green affordable housing is intended to uplift our dignity; it signals to us that we are worthy and part of the solution in caring for our planet.

Having a house or apartment you can afford to make your home is a defining moment for most anyone. When that place, your home, is green, it will not harm your health over time. You will be less likely to lose your home, because a green home reduces the costs of renting and homeownership and has been designed and built to address the impacts of climate change. Your green home will not isolate you from opportunities to live your best life.

The defining characteristic of green affordable housing is the integrative design process. This means that instead of developing *for* people, we develop *with* people. Instead of developing *in* a community, we develop *with* a community. When we stop using the tools of community development to wield power and influence over people, we can imagine a better place together.

Community development can no longer take a paternalistic approach to developing communities by imposing values that developers believe will be good for the community. It is time to stop experimenting on communities through our housing, which is arguably the greatest determinant of our health and life course. This happens every day in America when we build and retrofit affordable housing without asking what residents need. Part of the ripple effect from green affordable housing is that we can have high-performing housing from an energy and water perspective and from a human one. I am reminded of Bernice Aquino in Baltimore. She wanted to make her home healthier for her daughter, Jacquirelyn. To do so, she had to let people into her home, people who could tell her everything that was contributing to her daughter's ill health. This could have been a humiliating process. However, when Bernice invited me into her home, she was the one who showed me the improvements and what a difference each one made and why. She is now educating her friends and neighbors who are experiencing similar situations in their homes. It was not an extractive process of taking away her dignity by fixing everything according to the contractors' ideas and preferences. Through green retrofits we have the opportunity to stop "green grabbing" and taking opportunities to do what we think should be done to a person's home in the name of the environment. If we are fortunate enough to be invited into someone's home, our purpose is to listen to what they need or want and, through conversation with them, suggest pathways for getting there.

A green affordable housing project that was designed with the residents of that community, who asked for healthy housing, is Via Verde[1] in the Bronx, New York. The development includes a green roof, which provides a respite for the residents from stress and the bustle of city life, helps to absorb noise and cool the ambient air during warm weather, and lowers the heat-island effect in the neighborhood. One elegant gesture, multiple benefits.

Center Racial Justice

Without first acknowledging that our housing and development patterns reflect historical practices and policies that segregated communities on the basis of race, we cannot proactively target housing rehabilitation and production in communities that have been left behind because they are where the Black, Indigenous, people of color, and poor White people have lived. Investments in housing must be commensurate with investments to ensure that people living in the housing have access to transportation options, schools, grocery stores, and other necessities. To green all the housing in this country, racial justice must be at the center. We will only be successful if we lead the effort with antiracist organizations, land use and zoning, building standards, programs, policies, funding, and financing.

Sunshine Mathon, executive director of the Piedmont Housing Alliance in Charlottesville, Virginia, is a former Enterprise Rose Fellow with Foundation Communities. He has led some of the most advanced green affordable housing developments in our country. In August 2017, after he had been in his position for less than 100 days, White supremacist and neo-Nazis protesters rallied in opposition to the removal of Confederate monuments and memorials in Charlottesville. During the protest, a self-identified White supremacist deliberately drove into a crowd of counterprotesters, killing a woman named Heather Heyer and injuring many others.

One of Piedmont's housing developments is three blocks from where Heather was killed. Friendship Court, once named Garrett Square, has a past that epitomizes the ills of urban renewal and its racist policies.[2] The housing had deteriorated over time since 1978, when it was developed, due to the lack of investment into its maintenance by the public housing authority. The Piedmont Housing Alliance, which became the managing partner for the development in 2002, is actively working with the community to redevelop the housing while not displacing the current residents. The residents of Friendship Court elected nine residents as representatives of the community to codesign the redevelopment framework. Because of the recent activities, Piedmont Housing has begun viewing its own history and practices through a racial justice lens. Depending on what the residents decide, Friendship Court could become the first antiracist, green, affordable housing development in the country. They are actively pursuing this process. Friendship Court's state financing requirements dictate that it will be green. Sunshine will be reviewing every choice made in the redevelopment process, from the procurement of solar panels to the planting of trees, both to assess whether the choices are indeed antiracist and to avoid wherever possible polluting the planet and the people. With the patience and permission of all involved, Friendship Court could demonstrate the future of green affordable housing with racial justice inextricably rooted in its core.

"Because the systems in which we all operate have clear roots of construction in racist principles that advantage white people—the housing system being of particular note," shared Sunshine, "the act of ignoring these roots results in participation in and further upholding of the system itself. So either we are antiracist, actively working to dismantle the racist principles in the system, or we are racist, whether intentional or not."

By placing humans at the center of this work, we confront the past and our mistakes and intentional wrongdoings. We have not yet used the integrated design process—a core tenet of green housing development—to do this, but we must. We can begin to use the integrated design process to address the history of housing policies and finance to stop perpetuating their use to discriminate based on race and income.

In the green building movement, and Green Communities specifically, very little has been done to explicitly examine the effects of a gray housing system on Indigenous persons, Blacks, and persons of color. A report by the National Association for the Advancement of Colored People found that the six million people living within three miles of a coal-fired power plant had average per capita incomes of $18,400, and 39 percent were people of color and were predominantly Black.[3] Along with emitting greenhouse gases and accelerating climate change, pollution from coal plants is linked to health issues such as asthma attacks and heart disease.

Between 2007 and 2013, the net wealth of the median Black household fell from 10 percent to 8 percent of median White household wealth, largely the result of the differential impact of the Great Recession.[4] "White family wealth was seven times greater than Black family wealth and five times greater than Hispanic family wealth in 2016. Despite fluctuations over the past five decades, this disparity is as high as or higher than it was in 1963."[5] When the Federal Housing Authority guaranteed bank loans to developers, restricting them to building housing for only White households and requiring deeds to those homes that prevented their being resold to African Americans, it kick-started the acceleration of today's current racialized wealth gap.[6]

These disparities in wealth and income are a result of unconstitutional policies that have prevented Indigenous persons and people of color from acquiring transformative wealth.[7] We had housing policies in the middle of the 20th century that excluded African Americans from living in suburban communities. This forced Blacks and people of color into rental housing situations in areas where we then allowed polluters and extractive industries to locate, leading to deplorable conditions the likes of which caused outrage—and eventually protests—on what we now herald as Earth Day.[8] People living in areas with higher rates of poverty have worse outcomes across a range of social and economic measures. Native Americans have the highest rates of poverty of any population group in the United States, followed by African Americans and Latinx. From sidewalks to social capital, differences in safety and neighborhood quality lead to differences in outcomes, as the case studies jointly produced by the Community Affairs Offices of the Federal Reserve System and the Brookings Institution show.[9] Children in poor families that were able to use a voucher to move to a low-poverty neighborhood saw a 16 percent increase in college attendance and were much less likely to become single parents, compared to a control group; those who moved earned 31 percent more in their mid-20s.

Black people are 1.7 times more likely than the rest of the population to occupy homes with severe physical problems, not because they want to but because that is where they have been forced to live after decades of housing discrimination. Disparate investments in some communities but not others has caused widely varying neighborhood conditions that have a significant impact on the health of people who are paid low wages. Substandard housing conditions—such as pest infestation, lead paint, faulty plumbing, and overcrowding—disproportionately affect Indigenous persons and Black families and contribute to health problems, such as asthma, lead poisoning, heart disease, and neurological disorders. This is because our systems have allowed investment in some communities and not others. Concentrated housing inequity disproportionately exposes communities of color to environmental pollutants and isolates these populations from essential health resources, such as improved recreational spaces; quality pharmacies, clinics, and hospitals; and healthy food options. As Keeanga-Yamahtta Taylor shares in her book *Race for Profit*, "There has not been an instance in the last 100 years when the housing market has operated fairly without racial discrimination."[10]

Our country has a practice of drawing and redrawing maps in ways that favor White privilege and power. President Andrew Jackson urged Congress to pass the Indian Removal Act, which essentially redrew a map forcing eastern tribes to lands west of the Mississippi, laying the groundwork for Congress to pass the Indian Appropriations Act. This included 138 million acres promised to Indigenous people for their ownership and use. Over time 90 million of those acres have been "lost," by the federal government and what remains has been divided multiple times based on how the federal government mandated land to be passed down from one generation to the next. We did the same to Blacks by restricting them from the legal system, preventing access to services that could create recognized titles to land that they purchased. Land had to then be passed down without a will, and the land becomes legally categorized as heirs' property. Descendants receive only a fractional interest in part of the land. Without clear title, owners face structural barriers to access loans and federal programs that would allow for investment and upkeep of the land. The US Department of Agriculture has recognized it as the leading cause of Black involuntary land loss.[11]

The redlining practices continued the trend by preventing people of color and from certain religions from owning homes, which has had lasting consequences. We are a country that has accumulated wealth through people owning their homes and using their homes as collateral to access capital. Without owning a home or having title to your land, you have no way of accessing the capital markets other than by earning enough money to invest. Ibram X Kendi has drawn the line for all of us in his book *How to Be an Antiracist* by successfully arguing that we are either racist or antiracist—there is no middle ground. Without purposefully establishing antiracist housing policies and programs, including the Green Communities Criteria, we are perpetuating racist ones.

To ensure that our efforts to green all affordable housing do not perpetuate racial inequality, we must intentionally put racial justice front and center in all efforts to address the housing and climate crises. One way we can do this is to seriously implement the Affirmatively Furthering Fair Housing provision of the Fair Housing Act of 1968, which means

> taking meaningful actions, in addition to combating discrimination, that overcome patterns of segregation and foster inclusive communities free from barriers that restrict access to opportunity based on protected characteristics. Specifically, affirmatively furthering fair housing means taking meaningful actions that, taken together, address significant disparities in housing needs and in access to opportunity, replacing segregated living patterns with truly integrated and balanced living patterns, transforming racially and ethnically concentrated areas of poverty into areas of opportunity, and fostering and maintaining compliance with civil rights and fair housing laws. The duty to affirmatively further fair

housing extends to all of a program participant's activities and programs relating to housing and urban development.[12]

The Greater New Orleans Fair Housing Action Center is actively implementing the New Orleans Affirmatively Further Fair Housing plan by building coalitions to keep the city accountable and investigating violations.

In 2015, the US Department of Housing and Urban Development issued a new rule to affirmatively further fair housing, which included a requirement for certain participants in HUD's program to conduct an Assessment of Fair Housing. But in May 2018, it suspended this requirement. The City of Dallas continued with its assessment, which uncovered fair housing findings, including that the "nonwhite population and the population in poverty disproportionately reside in Dallas than in the region." Similarly, the rate of housing problems remains greater in Dallas than in the region: "Black and Hispanic households face housing problems and cost-burden challenges at a higher rate and with greater geographic dispersion than do white households," and "the data show an increasing level of nonwhite/white segregation characterized by clear spatial patterns."[13] These types of regional analyses are critical for seeing the overall picture of who has access to housing and how well that housing connects them to transportation options and employment opportunities. Because the research team was specifically looking at contributing factors to fair housing issues, the solutions can be rooted in overcoming discrimination, which, according to the study, "manifests in many compounding ways through community opposition, source of income discrimination, lending discrimination, and private discrimination, which tends to exacerbate housing challenges."

Again, it is not only about who gets to live in green housing; it is who is benefiting from the whole ecosystem involved in meeting this crisis. Who owns the manufacturing facilities for the inputs of all the materials and products used in the residential market? Who owns the businesses from which we are purchasing materials? Who owns and leads the design and engineering firms that develop plans for the housing? Who owns the land where the housing is developed or retrofitted? Who owns the housing? Who invested in the construction of the housing? The list goes on. These questions must be baked into our policies and financial institutions to ensure racially just answers if we want racially just outcomes.

Antiracist green affordable housing cannot fall on the shoulders of one community. We must all persistently demand it through our different roles, institutions, and levers of power. Perhaps the first place to start is by understanding who is making the decisions. When it comes to affordable housing and community development, if you examine the leadership of those sectors, it is predominantly White. Author Miriam Axel Lute wrote in *Shelterforce*,

> Neighbor Works reports that among its network membership, out of 247 CEOs/executive directors, 197, or 80 percent, are White. Nathaniel Wright, a former city planner and assistant professor of political science at Texas Tech University who studies Centers for Disease Control and Prevention outcomes, obtained similar numbers. In 2016, he surveyed a random sample of 1,000 of the 2,895 organizations coded on the GuideStar database as doing housing development, construction and management, housing rehabilitation, housing support, community and neighborhood development, economic development, and urban and community economic development. Of the 350 that responded, 84 percent had executive directors who were White, 11 percent African American, 5 percent Hispanic, 2 percent Native American, and 1 percent Asian American.[14]

I spoke with Joseph Kunkel, a citizen of the Northern Cheyenne Nation and director of the Sustainable Native Communities Design

Lab at Mass Design, about this statistic. He shared with me that there are approximately 50 Indigenous persons who are registered architects. The numbers are only slightly better for African Americans, Latinx, and Asian Americans. We cannot expect to advance antiracist green affordable housing when the designers, developers, and contractors are mostly White. We have a long way to go in the affordable housing and community development sectors to ensure that people in and from the community are in positions of power to make decisions.

However, this goes beyond the percentage of persons of color and women employed at organizations. Antiracism is about addressing our broken systems rooted in systemic racism and created to diminish the power of women and people of color. These systems are embedded in organizations and must be examined for real and lasting change to occur. This is at the heart of the protests following the murder of George Floyd by policemen in Minneapolis. Protestors are calling for the funding of police departments to be redistributed for housing, economic opportunities, and investments in communities. The training, weapons, and practices within most of our police departments are influenced by implicit bias and racial discrimination. Restructuring these departments goes deeper than just the face of a department or an organization, extending to its policies, procedures, and decision-making structures to prevent disproportionate outcomes that negatively affect Black lives.

Pursue Pathways toward Economic Justice

We know that green affordable housing can advance a clean energy transition for communities that have been left behind, squeezed out, and put on the frontlines of injustice. Part of that transition must include connecting the households to investment strategies that enable them to benefit financially even when they are renting.

We know that kids living in stable, affordable homes earn more as adults. We know that for every 100 affordable rental homes, $11.7 million is generated in local income, $2.2 million is generated in taxes and other revenue for the local government, and 161 local jobs are created (see Figure 5.1). Imagine if we met our immediate housing gap of 7.2 million affordable homes for families most in need. That would result in $842 billion in local income, $158 billion in taxes and other revenues for the local governments, and 11.5 million in local job creation.[15] As reported by Our Homes, Our Future, the GDP would be higher if families had better access to affordable housing.

In green homes, is there a way to capture some of the utility cost savings to also benefit the community? Groups like Elevate Energy in Illinois are proving that this is possible. If there is a community garden, can people living there sell produce? If there is a community room, can it include facilities for residents to make furniture or other products to sell? The way to real wealth creation is through having an equity share in the businesses that build and manufacture homes and in the property and commercial spaces that are included in the development. The community land trust model is one idea. Right to the City Alliance says on its website that it is "building a strong housing and urban justice movement in America and beyond. The idea of a right to the city frames and activates a new kind of urban politics that asserts that everyone, particularly

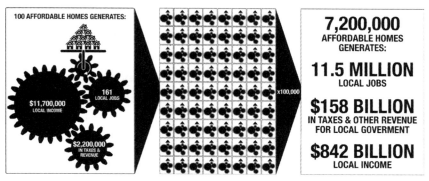

Figure 5.1. Local economic impact from affordable housing (Source: Our Homes, Our Voices)

the disenfranchised, not only has a right to the city, but as inhabitants, all have a right to shape it, design it, and operationalize an urban human rights agenda." They are working to advance a homes for all platform that "would put control back into the hands of the community and help to accumulate wealth and equity for the residents."

Green and Healthy Homes Initiative (GHHI) helped create a bill that passed in Maryland to eliminate lead poisoning while simultaneously creating at least 750 new "risk assessor" jobs. These positions were filled by persons from impacted communities who became trained in identifying lead hazards in homes. It is known that for each IQ point lost from lead exposure, there will be a loss of $17,815 in lifetime earnings compounded by lower hourly wages and greater health costs. The positions started at $55,000 a year. Through an effort to train and hire residents in communities with high rates of lead poisoning, this type of public investment led to wealth creation for impacted families, both directly and indirectly. Linwood Brown had experienced being unhoused prior to completing training sessions in risk assessment. He now works for GHHI, has a retirement account, health care insurance, paid vacation and sick days, and a family-sustaining salary that has supported college educations for his children.

Workers on green affordable housing developments are better off healthwise and possibly economically as well, given a study by Brookings that reports workers in clean energy earn higher and more equitable wages when compared to all workers nationally. Mean hourly wages exceed national averages by 8 to 19 percent. Clean energy economy wages are also more equitable; workers at lower ends of the income spectrum can earn $5–$10 more per hour than in other jobs.[16] Through reports by GHHI in their green jobs workforce development programs, such as the Center for Employment Opportunity in Buffalo, the average hourly wages increased by $2–$4 for GHHI workers who were cross-trained in lead hazard reduction, healthy homes, and weatherization skills and obtained various industry certifications—translating into annual earnings increases of $4,000–$8,000 for those individuals, although that wage too is still below what most need to afford decent housing in most cities.

Community solar is a promising pathway to transition communities to more clean sources of power while allowing residents to eventually own the systems and to benefit financially from lower electricity bills. In Washington, DC, local jurisdictions are incentivizing community solar through its Department of Energy and Environment and the Renewable Portfolio Standard Expansion Amendment Act of 2016. In DC, Jubilee Housing installed a community solar and storage system at its Maycroft Apartments, which provide housing to over 60 families, some of which had been homeless and others of which had with very low incomes. The community solar system credits each resident with a portion of the proceeds from the electricity generated by the system, which dramatically cuts their energy burden each month. The battery storage provides at least three days of power during an outage, allowing residents to have lighting, refrigeration for medications, and outlets for charging devices.

The real win will be to manufacture the parts and panels cleanly in communities and to ensure that entities in the communities have the opportunity to own and expand those manufacturing facilities. We must prepare for that reality now by cultivating local talent to acquire the skills, networks, and access to capital needed to grow our renewable energy industry.

Commit to Principles of Environmental and Climate Justice
The "inescapable network of mutuality" spoken about by the Reverend Martin Luther King Jr. includes Mother Earth. When we extract natural resources to build a home we must do so in a responsible and sustainable manner. If we are extracting from nature in ways that prevent the planet from flourishing, then neither will flourish. It does not matter if we are doing it in the name of providing a social good like affordable housing; we must stop and change our practices regardless.

The Environmental Protection Agency (EPA) defines *environmental justice* as "the fair treatment and meaningful involvement of all people regardless of race, color, national origin, or income with respect to the development, implementation, and enforcement of environmental laws, regulations, and policies." The EPA asserts that environmental justice "will be achieved when everyone enjoys the same degree of protection from environmental and health hazards and equal access to the decision-making process to have a healthy environment in which to live." But organizations involved in the pursuit of environmental justice argue that there is a simpler definition: "Stop poisoning people."

To achieve a future where we meet our housing goals with green afford-able housing for all, we must commit to these principles of environmental justice because even now, with current green building standards and a growing focus on healthy buildings, we are failing to look past the struc-ture. A common material metric in green building programs is to purchase materials within a 500-mile radius. What if instead the metric were that the production of that material caused no harm to people or the planet within a 500-mile of where the ingredients were sourced and the material manufactured? An environmentally just future can be measurably advanced by a commitment to green affordable housing if it is done with intention. Imagine the use of chemistry that does not result in classes of toxic chemi-cals but results in materials that earn the label of "clean energy."

Climate justice "insists on a shift from a discourse on greenhouse gases and melting ice caps into a civil rights movement with the people and communities most vulnerable to climate impacts at its heart," said Mary Robinson in 2019.[17] She was the first female president of Ireland and a former UN High Commissioner for Human Rights. None of the current green building standards, including the Green Communities Criteria, have gone far enough to advance climate justice.

Climate justice entails placing people who are most impacted by cli-mate inaction at the center of a transition to a world that stops exploiting nature and people for profit and power. If all we have are tens of millions of green homes that allow for business as usual to continue, we will not have realized the opportunity in front of us. Committing to green all the new and existing housing provides us with a platform to stop extracting resources at the expense of the planet and the people in those communities. This includes extracting the financial resources from communities instead of enabling the generation of wealth that provides for long-term affordability and stability. People in need of affordable housing must be centered in and driving the solutions so that they themselves benefit.

At the very least, we must halt all new expansions of our fossil fuel infrastructure, which are predominantly proposed to run through communities that are seen as having little power or influence. Communities engaged in climate justice efforts, however, are proving that they do have power. We witnessed this at Standing Rock and in Union Hill, Virginia, where the community leveraged their network to stand up against the governor and the Virginia Air Pollution Control Board. Dominion Energy was seeking permits to locate a compressor station in Union Hill, a historic African American community settled by freedmen and freedwomen after the Civil War. The Fourth Circuit of the US Court of Appeals sided with the community. While we pursue climate justice through clean energy, forest conservation, and green housing, we must also defend against the injustices of powerful corporations trying to grab land and air rights from the places and people they perceive as being weak and voiceless to lay obsolete infrastructure at odds with our climate action targets, sustainable development goals, and basic human rights.

Conclusion

We need a complete systems change and to put the power back into the hands of the people demanding access to green housing that they can afford. Within the housing sector, power rests in many places but not with the millions of people who are cost burdened and locked out of safe and dignified green housing. Green Communities has gone a long way toward providing a platform that puts the voice of the resident back into the design and decision-making process, but we still have a long way to go.

The Green New Deal is a global concept to address climate change by 2030, with massive public sector investments. In the United States, the Green New Deal resolution was introduced to the US Congress by Representative Alexandria Ocasio-Cortez of New York and Senator Edward J. Markey of Massachusetts. As a resolution, it provides a framework to guide a host of policies that would be needed to actually get us to a 100 percent clean and renewable energy economy. However, it has ignited cities, regions, states, and Indigenous communities to offer up additions, and localized versions. Putting any of these ideas into action in ways that allow for a just transition will take a massive infusion of public dollars. Although meeting our housing crisis with green affordable housing would not achieve everything referenced in the Green New Deal, it must be part of it and not an afterthought—and not just the energy efficiency elements but the holistic approach, as set forth in the Green Communities Criteria.

We have a historic opportunity to start this work first in the communities we have systematically underserved or that, when we have invested there, we have done great damage with infrastructure projects that have decimated communities and natural systems alike. I ask myself, Why are we so unwilling to meet our housing demand? My reaction is similar to that of Greta Thunberg, the young climate activist. She spoke to the UN Climate Action Summit in 2019 and said, "For more than 30 years, the science has been crystal clear. How dare you continue to look away and come here saying that you're doing enough when the politics and solutions needed are still nowhere in sight. You say you hear us, and that you understand the urgency, but no matter how sad and angry I am, I do not want to believe that. Because if you really understood the situation and still kept on failing to act then you would be evil and that I refuse to believe." It is difficult to reconcile how good people can allow other people not to have a dignified place to live. Now, with the impacts of climate change, it is the people without a place to live who will quite possibly suffer the most.

We can no longer focus solely on the physical process of putting a roof over someone's head without also considering the social and cultural context of where that housing will be situated within the community's fabric and how it will perform to support people looking for access to opportunity. Green housing that is affordable to all of us provides a platform to weave these connected issues together. It can also aggregate demand for cleaner and healthier products manufactured locally with family-sustaining wages that assist a circular economy where communities are made greener, not more gray, in the process.

As Bryon Stevenson, the founder of the Equal Justice Initiative and author of *Just Mercy* asserts, the opposite of poverty is not wealth, it is justice. When we think about meeting our housing crisis, it is not just a matter of building millions of shelters or allowing people to sleep in the parks or cranking out cheap housing in places where nothing else exists. Meeting the demand for housing at a scale that would address the crisis is not a simple endeavor. It is a complex challenge, but agreeing to do something about it is simple. We must put the full resources of our collective action behind solving it in a way that leaves all communities better off, in a way that furthers justice. Anything less is unacceptable.

Conclusion

What is the use of a house if you haven't got a tolerable planet to put it on?

—Henry David Thoreau

IF WE DO NOTHING TO ADDRESS THE housing and climate crises, living conditions in this country will get worse for all of us. Lucky for us, we know how to take action, and by doing so, we will learn more and better ways to mitigate both crises in ways we may not have considered possible. It is a tremendous opportunity in front of us, full of untapped potential, and one that we can achieve within a decade if we make the commitment to do so. This current time of exploiting the planet and workers must end. The climate crisis is a stark reminder that people and the planet have limits. We are tired of being squeezed, left out, unseen, and extracted from to build wealth for an ever-shrinking few. We are ready for a just transition to a clean energy future, one where everyone has housing they can afford, regardless of race or income. It is time to think differently about what is possible and to take action to make it so.

The green housing movement recognizes that the desire to conquer our environment has led to really crappy living experiences. Just because we can heat or cool any structure, at any time of day, at any cost, anywhere in the country does not mean that we should. Green housing extends the sensibility that we must listen to the people and to the environment when we are building and rehabilitating houses. This is the fundamental underpinning of the Green Communities Criteria that must still be fully realized.

Since Enterprise launched the Green Communities Criteria in 2004, our understanding of the benefits of green affordable housing has progressed. We have begun to think differently about how we design, locate, construct, rehabilitate, operate, and maintain housing. We now know definitively how to deliver health, economic, and environmental benefits through green methods and materials. We can measure the long-term savings to residents, society, and the planet when we invest in green affordable housing instead of gray housing. There are examples of green affordable housing across the country that have met and exceeded the Criteria.[1] These homes are available to people with very low incomes and persons who were formerly unhoused. They meet passive house standards, are net-zero energy and net-zero water, and have integrated methods to prepare the housing for extreme weather events. It is no longer a question of how, it is a question of when. When will we provide housing at the pace and scale needed to ensure that everyone has access to a green home they can afford? When are we going to invest in the preservation and delivery of housing in this country in a way that signals our desire for everyone and the planet to flourish?

There is an elegance and a harmony to communities that have committed to green housing. The housing connects to the community around it, people have transportation options, they have access to green space, rainwater is being put to use, and increasingly the sun and wind are being harvested to provide electricity to homes in the development, and often to surrounding homes, in ways that add to the financial security of residents.

To galvanize the support needed to unlock the required resources for the environmental and social justice movements, we must recognize the tremendous opportunity of doing so in environmentally beneficial ways while delivering on a fundamental human right, without which there is no justice.

When I was living in my sixth home at the age of 10 after moving from South East England to Southern California, I began an effort to save ice cubes. It sounds crazy now, and it was crazy then, but I was struck by my family's habit of throwing once-used ice cubes into the sink to melt into the drain or emptying large coolers of ice onto the driveway when the brown grass was only inches away. This caught my attention because of the forest fires in the mountains that were visible to me from our backyard. No one in my family did anything other than tease me when I asked them to put their ice cubes back in the freezer or throw them into the plants or onto the grass. I knew the drought was causing the fires, and I was willing to reach for whatever I could to try and help conserve water. That, however, was a laughable solution, completely out of scale with the problem. We do not have time to complacently further solutions to address climate change that are not on scale with the challenge. We need holistic, comprehensive efforts that set us on a successful path toward the transition to a clean energy future by ensuring that everyone has a place to call home and, in so doing, realize multiple other benefits.

This must be our moonshot. Let us stop acting like we do not have housing and climate crises. We know we are capable of meeting these interconnected challenges, and although we may not have all the solutions, we have enough to get started. The following are a few. This is by no means an exhaustive or detailed list, but perhaps it can help point the way forward:

1. **Make a national commitment**. Create a national commitment to ensure over the next decade that everyone has access to a house they can afford regardless of race or income on a planet that is

thriving. This commitment would include the budget appropriations, incentives, and regulations necessary to assemble the level of human and financial resources needed from across sectors to be successful.

2. **Install a national green housing conductor**. Create a corresponding entity to the national commitment with the decision-making and budgetary authority behind it to track and account for progress on all fronts while identifying areas for acceleration and innovation. It would create an umbrella for the many siloed agencies, programs, studies, and centers focused on various aspects of our housing and climate action efforts to further coordination and opportunities for collaboration.

3. **Curb pollution**. Hold government accountable to regulate polluters of our air, water, soil, and bodies and stop the permitting to build new or expand existing facilities known to cause harm in communities and contribute to climate change. Require manufacturers of all products used for housing construction, rehabilitation, operations, and maintenance to make a Health Product Declaration® for each product. Invest in green chemistry and product development to build an industry that can meet our housing construction demands through healthy and environmentally beneficial products. Let us start asking and demanding to know what is in the products used to make our homes and those that we bring into our homes. We did it with food labels; we can do it with housing products. The ingredients of both end up in our bodies.

4. **Preserve and improve existing housing**. Stop the loss of existing affordable housing by assembling investment vehicles to acquire the properties, refinance them, and deliver green rehabilitation without displacement of current residents. These vehicles will have community participation and will result in

ownership structures agreed upon by the residents and community, including opportunities for community land trusts or other community-endorsed models. This would include vehicles to support the millions of small landlords who own one or two rental properties but as such are completely outside of the housing system because they do not access housing programs or accept renters with housing vouchers but are nonetheless important stakeholders.

5. **Train and fund a green housing workforce**. Fully fund training programs that result in a job force ready to construct new housing and deliver green rehabilitations of all existing housing, with a preference for community-based, minority- and women-owned businesses. This includes an opportunity for employment related to all green infrastructure installations, operations, and maintenance from solar systems to bioswales.

6. **Hold people in power accountable**. Use the power of our vote to ensure that all elected and appointed persons in positions of influence and power, including public utility commissions, are committed to and show action toward ending our housing and climate crises together in ways that advance racial, economic, and environmental justice. A single accountability framework will make known and rate the voting and decision-making records measured against housing and climate goals.

These action items may seem beyond your control. They are not. As Congressman John Lewis said in his address to participants of a NEWHAB meeting in 2017, "This is our planet. We all must learn to live together in a clean and safe environment. If not we will perish as fools. Stand up, speak up and speak out. When you see something that is not right, not fair, not just, we all have an obligation to do something about it."[2]

Remember Tracey, Bernice, and Max. Tracey was simply looking for a place she could afford to rent. Ask the elected officials in your community who approve zoning and land use decisions if new developments they approve will be affordable to people in your community that need housing. Ask what green building standard it will meet and if it will reduce asthma triggers as well as carbon emissions. Let those in positions of decision-making authority know if their answers are acceptable to you. Talk about it with your friends. Seek out networks such as NEWHAB for support. Like Tracey, we should be able to rent housing we can afford without having to oversee its development to ensure it will be healthy and affordable to operate and maintain. The developer should be held accountable to the community where they are delivering green housing so that Tracey can move in and have the security of knowing her home is healthy and affordable to maintain.

Additionally, we should have the security of knowing our communities will not be inheriting housing developments that will be polluting. Like Bernice do you know how green your own home is and where your electricity comes from? Find out. If you do not like the answers, call up GHHI, Elevate Energy, or an entity in the Relay Network and find out what you can do about it. They can identify someone to speak with in your community.

Lastly, we all have the power that Max tapped into—the power to organize. We can organize our friends and neighbors and demand better for our communities. You do not know where to start? Start by perusing the websites of any of the organizations mentioned in this book, from the Right to the City Alliance to the Center for Heirs' Property Preservation to the Sunrise Movement, and look to efforts like People's Action's Homes Guarantee to find out what they are doing in your community or ask how to get something started in your community. You can influence new housing development, you can retrofit your own housing, and you can organize your community or join organizers in your community to demand action on the housing and climate crises.

Over this decade we can finally deliver on the recognized human right to housing and avert catastrophic global warming while advancing racial, economic, and environmental justice. To do so will take fully funding all programs related to improving and preserving our existing housing stock, from weatherization programs to lead remediation to climate resilience and preparedness. We must use every tool available from every corner of the public, private, and nonprofit sectors to construct and rehabilitate housing that will be affordable and accessible to everyone, including to those with very low and no incomes, while keeping us all safe and healthy.

The science has shown us that our gray housing practices are making us sick, and our reliance on fossil fuels is choking our planet and causing global warming. Comprehensive and holistic green housing is achievable and is proven to reduce carbon emissions while boosting health and economic outcomes. Thank you to the thousands of organizers, residents, owners, developers, designers, planners, engineers, contractors, funders, investors, policy makers, and administrators who have already committed to making all affordable housing green. Let us continue innovating to drive down the price and increase the effectiveness of green materials, renewable energy and battery storage, and efficient appliances and mechanical systems while driving technology and closing the digital divide to allow for even greater monitoring and performance verification. The health and well-being of each person in this country and the future of our planet depend on it. Let us act like this is true, if not for us then for our children and our children's children.

The opportunity to change the course of human history truly is within our grasp.

Notes

Preface

1. Eric Tars, "Housing as a Human Right," *2017 Advocates Guide*, National Low Income Housing Coalition, https://nlihc.org/sites/default/files/AG -2018/Ch01-S06_Housing-Human-Right_2018.pdf.
2. *Nation under Siege*, Architecture 2030, accessed June 28, 2020, https:// architecture2030.org/nation_under_siege/.
3. Joseph G. Allen and John D. Macomber, *Healthy Buildings: How Indoor Spaces Drive Performance and Productivity* (Cambridge, MA: Harvard University Press, 2020).
4. Tracey Woodruff, "EPA Method Will Curtail Science Used in Chemical Evaluations," Program on Reproductive Health and the Environment, August 17, 2018, https://prheucsf.blog/2018/08/17/epa-method-will -curtail-science-used-in-chemical-evaluations/.
5. David Roberts, "As Coal Companies Sink into Bankruptcy, Who Will Pay to Clean Up Their Old Mines?," Vox, September 2, 2016, https://www .vox.com/2016/9/2/12757074/coal-bankruptcy-mine-cleanup.
6. "Does Climate Change Affect the Transmission of Coronavirus?," Harvard T. H. Chan School of Public Health, Center for Climate, Health, and the Global Environment, accessed June 28, 2020, https://www .hsph.harvard.edu/c-change/news/coronavirus-climate-change-and-the -environment/.

Introduction

1. Lisa Friedman, "What Is the Green New Deal? A Climate Proposal, Explained," *New York Times*, February 21, 2019, https://www.nytimes .com/2019/02/21/climate/green-new-deal-questions-answers.html.

2. See https://www.sunrisemovement.org/about/.

3. Joint Center for Housing Studies of Harvard University, *America's Rental Housing 2020*, 2020, https://www.jchs.harvard.edu/sites/default/files/Harvard_JCHS_Americas_Rental_Housing_2020.pdf.

4. Joint Center for Housing Studies of Harvard University, *The State of the Nation's Housing 2019*, 2019, https://www.jchs.harvard.edu/state-nations-housing-2019; see also https://www.jchs.harvard.edu/state-nations-housing-2019.

5. Baltimore is home to Pimlico racetrack, most famous for hosting the Preakness Stakes, and when a winning bet is placed on who will come in first, second, and third, it is referred to as a trifecta.

6. Diana Hernández, "Housing as a Platform for Health and Equity: Evidence and Future Directions," *American Journal of Public Health* 109 (2019): 1363–1366, https://ajph.aphapublications.org/doi/10.2105/AJPH.2019.305210.

7. "What Is the Circular Economy?," Ellen MacArthur Foundation, accessed June 24, 2020, https://www.ellenmacarthurfoundation.org/circular-economy/what-is-the-circular-economy.

8. Kim Seong Lee, "The Girl Who Silenced the World for 5 Minutes," YouTube video, April 18, 2008, 6:42, https://www.youtube.com/watch?v=TQmz6Rbpnu0. Twelve-year-old Severn Cullis-Suzuki spoke at the UN Conference on Environment and Development in Rio de Janeiro in June 1992.

9. Greta Thunberg, *No One Is Too Small to Make a Difference* (London: Penguin Books, 2019), 96.

Chapter 1. The Problem with Gray

1. Jung Hyun Choi, "Three Differences between Black and White Homeownership That Add to the Housing Wealth Gap," Urban Institute, February 28, 2019, https://www.urban.org/urban-wire/three-differences-between-black-and-white-homeownership-add-housing-wealth-gap.

2. Eric Tars, "Housing as a Human Right," National Low Income Housing Coalition, https://nlihc.org/sites/default/files/AG-2018/Ch01-S06_Housing-Human-Right_2018.pdf.

3. Jennifer Rudden, "Number of Owner Occupied Housing Units in the United States from 1975 to 2019," Statista, May 12, 2020, https://www

.statista.com/statistics/187576/housing-units-occupied-by-owner-in-the -us-since-1975.

4. Meghan Henry, Anna Mahathey, Tyler Morrill, Anna Robinson, Azim Shivji, and Rian Watt, Abt Associates, *The 2018 Annual Homeless Assessment Report to Congress*, US Department of Housing and Urban Development, December 2018, https://files.hudexchange.info/resources/ documents/2018-AHAR-Part-1.pdf.

5. "Ending Veteran Homelessness in Virginia: A Statewide Collaboration," National Alliance to End Homelessness, accessed June 23, 2020, https:// endhomelessness.org/resource/ending-veteran-homelessness-virginia -statewide-collaboration/.

6. Benjamin Applebaum, "America's Cities Could House Everyone If They Choose To," *New York Times*, May 15, 2020, https://www.nytimes.com/ 2020/05/15/opinion/sunday/homeless-crisis-affordable-housing-cities.html.

7. Elizabeth La Jeunesse, Alexander Hermann, Daniel McCue, and Jonathan Spader, *Documenting the Long-Run Decline in Low-Cost Rental Units in the US by State*, Joint Center for Housing Studies of Harvard University, September 2019, https://www.jchs.harvard.edu/sites/default/files/harvard _jchs_loss_of_low_cost_rental_housing_la_jeunesse_2019_0.pdf.

8. T. Wang, "National Price Index for Water and Sewerage Maintenance in the US from 2000 to 2018," Statista, March 12, 2020, https://www .statista.com/statistics/208519/price-index-for-us-water-and-sewerage -maintenance/.

9. California Public Utilities Commission, *CPUC Works to Help Mitigate Higher Energy Bills Utility Customers May Receive Due to Shelter at Home*, April 2, 2020, https://docs.cpuc.ca.gov/PublishedDocs/Published/G000/ M331/K358/331358410.PDF.

10. Energy Efficiency for All, "Living at Park's Edge," YouTube video, August 15, 2018, https://youtu.be/KSfo6jXxSSI.

11. Environmental and Climate Justice Program, NAACP, *Lights Out in the Cold: Reforming Utility Shut-Off Policies as If Human Rights Matter*, March 2017, https://www.naacp.org/climate-justice-resources/lights-out-in-the-cold/.

12. To learn more about Energy Efficiency for All and NEWHAB, its national social impact and learning network open to all individuals working collectively toward making affordable multifamily homes energy and water efficient, see the web page https://www.energyefficiencyforall.org/about/.

13. Jamie Ducharme and Elijah Wolfson, "Your Zip Code Might Determine How Long You Live—and the Difference Could Be Decades," *Time*, June 17, 2019, https://time.com/5608268/zip-code-health/.

14. Nadine Burke Harris, "How Childhood Trauma Affects Health across a Lifetime," filmed September 2014 at TEDMED, https://www.ted.com/speakers/nadine_burke_harris_1; Nadine Burke Harris, *The Deepest Well: Healing the Long-Term Effects of Childhood Adversity* (Boston: Houghton Mifflin Harcourt, 2018).

15. Mona Hanna-Attisha, *What the Eyes Don't See* (New York: Penguin Random House, 2019).

16. Xiao Wu, Rachel C. Nethery, M. Benjamin Sabath, Danielle Braun, and Francesca Dominici, "COVID-19 PM2.5: A National Study on Long-Term Exposure to Air Pollution and COVID-19 Mortality in the United States," Harvard University, April 24, 2020, https://projects.iq.harvard.edu/covid-pm.

17. *Particulate matter* is the term for a mixture of solid particles and liquid droplets found in the air. "Particulate Matter (PM) Pollution," US Environmental Protection Agency, last updated November 14, 2018, https://www.epa.gov/pm-pollution/particulate-matter-pm-basics.

18. Indigenous Environmental Network, Little Village Environmental Justice Organization, and National Association for the Advancement of Colored People, *Coal Blooded: Putting Profits before People*, April 2016, https://www.naacp.org/wp-content/uploads/2016/04/CoalBlooded.pdf.

19. For more information on environmental justice, see Audrea Lim, *The World We Need: Stories and Lessons from America's Unsung Environmental Movement* (New York: New Press, 2021).

20. Indigenous Environmental Network, Little Village Environmental Justice Organization, and National Association for the Advancement of Colored People, *Coal Blooded*.

21. Jim Erickson, "Targeting Minority, Low-Income Neighborhoods for Hazardous Waste Sites," *Michigan News*, January 19, 2016, https://news.umich.edu/targeting-minority-low-income-neighborhoods-for-hazardous-waste-sites/.

22. Amanda Willis, "The Lowdown on Landfills," Earth 911, March 30, 2009, https://earth911.com/eco-tech/the-lowdown-on-landfills/.

23. "Sources of Greenhouse Gas Emissions," US Environmental Protection Agency, accessed September 21, 2020, https://www.epa.gov/ghgemissions/sources-greenhouse-gas-emissions.

24. Barbara J. Lipman, *A Heavy Load*, Center for Housing Policy, October 2006, http://www.reconnectingamerica.org/assets/Uploads/pubheavyload1006.pdf.

25. "The Center for Neighborhood Technology's Housing and Transportation Affordability Index," Center for Neighborhood Technology, https://htaindex.cnt.org/about/.

26. "Most Recent National Asthma Data," Centers for Disease Control and Prevention, last reviewed March 24, 2020, https://www.cdc.gov/asthma/most_recent_national_asthma_data.htm.

27. Elizabeth Grossman, "Banned in Europe, Safe in the US," *Ensia*, June 9, 2014, https://ensia.com/features/banned-in-europe-safe-in-the-u-s/.

28. "Making Affordable Multifamily Housing More Energy Efficient: A Guide to Healthier Upgrade Materials," Energy Efficiency for All, September 2018, https://s3.amazonaws.com/hbnweb.dev/uploads/files/Qj5q/NRDC-3084%20Guide%20to%20Healthier%20Retrofit_Final.pdf.

29. Vytenis Babrauskas, Donald Lucas, David Eisenberg, Veena Singla, Michel Dedeo, and Arlene Blum, "Flame Retardants in Building Insulation: A Case for Re-evaluating Building Codes," *Building Research & Information*, 40, no. 6 (2012): 738–755, https://www.tandfonline.com/doi/abs/10.1080/09613218.2012.744533.

30. Jessica Leung, "Decarbonizing U.S. Buildings," Center for Climate and Energy Solutions, July 2018, https://www.c2es.org/document/decarbonizing-u-s-buildings/.

31. "Energy and the Environment Explained," US Energy Information Administration, June 2019, https://www.eia.gov/energyexplained/energy-and-the-environment/where-greenhouse-gases-come-from.php.

Chapter 2. The Promise of Green

1. For more information on community land trusts, see https://groundedsolutions.org/strengthening-neighborhoods/community-land-trusts.

2. "Location and Green Building," US Environmental Protection Agency, last updated September 21, 2017, https://www.epa.gov/smartgrowth/location-and-green-building.

3. Jacob Kriss, "What Is Green Building?," United States Green Building Council, August 6, 2014, https://www.usgbc.org/articles/what-green-building.

4. W. J. Fisk, D. R. Black, and G. Brunner, *Benefits and Costs of Improved IEQ in U.S. Offices*, Ernest Orlando Lawrence Berkeley National Laboratory, April 2011, https://iaqscience.lbl.gov/sites/all/files/benefits-1.pdf.

5. "'Healthy' Buildings Can Improve Workers' Performance," Harvard T. H. Chan School of Public Health, 2017, https://www.hsph.harvard.edu/news/hsph-in-the-news/healthy-buildings-can-improve-workers-performance/.

6. Collen Walsh, "How Masks and Buildings Can Be Barriers to the Coronavirus," *Harvard Gazette*, April 7, 2020, https://news.harvard.edu/gazette/story/2020/04/how-buildings-masks-can-be-barriers-to-coronavirus/.

7. For more information on the Living Building Challenge, see https://living-future.org/.

8. Robert B. Peña, *Living Proof: The Bullitt Center*, Center for Integrated Design, University of Washington, 2014, 40, accessed August 16, 2020, http://www.bullittcenter.org/wp-content/uploads/2015/08/living-proof-bullitt-center-case-study.pdf.

9. The Rose Fellowship partners emerging architectural designers and socially engaged arts and cultural practitioners with local community development organizations to facilitate an inclusive approach to development that results in green, sustainable, and affordable communities. "Rose Fellowship," Enterprise Community Partners, accessed June 23, 2020, https://www.enterprisecommunity.org/solutions-and-innovation/rose-fellowship.

10. "2020 Enterprise Green CommunitiesSM Criteria," Enterprise Green CommunitiesSM, accessed June 23, 2020, https://www.greencommunitiesonline.org/.

11. Carrie Niemy, "Big Idea: Always Question Your Assumptions," Enterprise Green Communities, October 7, 2019, https://www.enterprisecommunity.org/blog/big-idea-always-question-your-assumptions.

12. Common Sense and Boston Consulting Group, *Closing the K–12 Digital Divide in the Age of Distance Learning*, 2020, https://www.commonsensemedia.org/sites/default/files/uploads/pdfs/common_sense_media_report_final_7_1_3pm_web.pdf.

13. The small team included myself, Stephen Goldsmith, Gail Vittori, Dalila Brooks, and Doris Koo.

14. For example, Nihls Bohlin, the person most credited with the modern seat belt, had developed ejection seats for fighter aircraft in the 1950s. He knew how to keep people safely strapped into moving vehicles. He made a compelling case to car manufacturers to buy into the additional expense of installing the safety device because he knew what he was talking about.

15. "Cool Roofs," US Department of Energy, https://www.energy.gov/energysaver/design/energy-efficient-home-design/cool-roofs.

16. The Green Communities Criteria serves as a platform that weaves connected issues together. This community-centered way of thinking is what helped us to make the case to Congress. The House of Representatives passed a landmark green affordable housing bill with wide bipartisan support, the HOPE VI Improvement and Reauthorization Act of 2007 (H.R. 3524). It was created to provide $800 million annually from 2008–2013 for mixed-income communities that incorporate the Green Communities Criteria. This was the first time the House had passed a bill authorizing comprehensive environmental principles in a major housing program.

Chapter 3. Learning from the Green Communities Criteria

1. Dana Bourland, *Incremental Cost Measurable Savings* (Columbia, MD: Enterprise Community Partners, 2009).

2. Washington State Occupational Respiratory Disease Program, *Isocyanate-Based Foam and Work-Related Asthma*, 2017, http://www.lni.wa.gov/safety/research/files/42_03_2017_isocyanate_foam.pdf.

3. The term *additionality* refers to those carbon emissions that would not have been achieved but for the funding provided through the sale of offset. The Kyoto Protocol refers to "additionality" as reductions in emissions that are additional to any that would occur in the absence of the certified project activity.

4. See the Gold Standard website, https://www.goldstandard.org/.

5. The Green Communities Offset Fund no longer exists. It bought additional carbon emissions reductions from affordable housing developers at twenty dollars per ton for five years. This is a model that could be brought back to the market with an investment to cover the operating and verification costs.

6. "First Ever Carbon Offsets Sale Aids Development of Green Affordable Housing," Market Watch, June 30, 2018, http://www.ipedconference

.com/powerpoints/First_Ever_Carbon_Offsets_Sale_Aids_Development
_of_Green_Affordable_Housing.pdf.

7. "2020 Enterprise Green Communities Criteria," Enterprise Green Communities, accessed June 23, 2020, https://www.greencommunitiesonline.org/.

8. "Five Reasons Why Climate Change and Toxic Chemicals Are Connected," Healthy Building Network, December 2019, https://healthybuilding .net/blog/533-five-reasons-why-climate-change-and-toxic-chemicals-are -connected.

Chapter 4. The Challenges to Greening Affordable Housing for All

1. National Association of State Energy Officials and Energy Futures Initiative, *The 2020 U.S. Energy & Employment Report*, 2020, https://www .usenergyjobs.org/.

2. Ariane Hegewisch, Jeff Hayes, Tonia Bui, and Anlan Zhang, *Quality Employment for Women in the Green Economy* (Washington, DC: Institute for Women's Policy Research, 2013).

3. Jonathan Watts, "Surge in Chemical Use 'a Threat to Health and Environment,'" *Guardian*, March 12, 2019, https://www.theguardian.com/environment/2019/ mar/12/surge-in-chemical-use-a-threat-to-health-and-environment.

4. "The Cost of Affordable Housing: Does It Pencil Out?," Urban Institute in Partnership with the National Housing Conference, July 2016, https:// apps.urban.org/features/cost-of-affordable-housing/.

5. "Trolley Square," Building Green Gallery, accessed June 28, 2020, https:// web.njit.edu/abs/green/trolley_square.htm.

6. Maya Brenan, "To Reduce Government Costs, Spend Wisely on Housing," Housing Matters, December 14, 2016, https://housingmatters .urban.org/articles/reduce-government-costs-spend-wisely-housing.

7. Jon Braman, Steven Kolberg, and Jeff Perlman, *Energy and Water Savings in Multifamily Retrofits*, Stewards of Affordable Housing for the Future, June 2014, https://sahfnet.org/sites/default/files/uploads/resources/multifamily _retrofit_report.pdf.

8. See https://relaynetwork.org/about/.

9. "Fact Sheet: Housing First," National Alliance to End Homelessness, April 2016, http://endhomelessness.org/wp-content/uploads/2016/04/housing -first-fact-sheet.pdf.

Chapter 5. A Just Future

1. "Via Verde," Jonathan Rose Companies, accessed June 23, 2020, http://www.rosecompanies.com/projects/via-verde/.

2. Jordy Yager, "The Reimagining of Friendship Court," Charlottesville Tomorrow, accessed June 24, 2020, https://www.cvilletomorrow.org/specials/friendship-court/#.

3. Indigenous Environmental Network, Little Village Environmental Justice Organization, and National Association for the Advancement of Colored People, *Coal Blooded: Putting Profits Before People*, April 2016, https://www.naacp.org/wp-content/uploads/2016/04/CoalBlooded.pdf.

4. Dayna Bowen Matthew, Edward Rodrigue, and Richard V. Reeves, "Time for Justice: Tackling Race Inequalities in Health and Housing," Brookings, October 19, 2016, https://www.brookings.edu/research/time-for-justice-tackling-race-inequalities-in-health-and-housing/.

5. "Nine Charts about Wealth Inequality in America," Urban Institute, last updated October 5, 2017, https://apps.urban.org/features/wealth-inequality-charts/.

6. Richard Rothstein, *Color of Law* (New York: Liveright, 2017).

7. john a. powell, *Racing to Justice* (Bloomington: Indiana University Press, 2012).

8. "This History of Earth Day," Earth Day Network, accessed June 24, 2020, https://www.earthday.org/history/.

9. Federal Reserve System and the Metropolitan Policy Program, Brookings Institution, *The Enduring Challenge of Concentrated Poverty in America: Case Studies from Communities across the U.S.*, 2008, https://www.brookings.edu/wp-content/uploads/2016/06/1024_concentrated_poverty.pdf.

10. Keeanga-Yamahtta Taylor, *Race for Profit: How Banks and the Real Estate Industry Undermined Black Homeownership* (Chapel Hill: University of North Carolina Press, 2019), 261.

11. Lizzie Presser, "Their Family Bought Land One Generation after Slavery. The Reels Brothers Spent Eight Years in Jail for Refusing to Leave It," ProPublica, July 15, 2019, https://features.propublica.org/black-land-loss/heirs-property-rights-why-black-families-lose-land-south/?utm_source=pardot&utm_medium=email&utm_campaign=majorinvestigations.

12. "Affirmatively Furthering Fair Housing," HUD Exchange, July 16, 2015, https://www.federalregister.gov/documents/2015/07/16/2015-17032/affirmatively-furthering-fair-housing.
13. "City of Dallas Draft Assessment of Fair Housing Report," North Texas Regional Housing Assessment, accessed June 13, 2020, https://ntrha.uta.edu/.
14. Miriam Axel-Lute, "Who Will Lead Community Development Corporations," *Shelterforce*, August 23, 2017, https://shelterforce.org/2017/08/23/who-will-lead-community-development-corporations/.
15. "Housing Is Infrastructure," National Council of State Housing Agencies, February 2018, https://www.ncsha.org/wp-content/uploads/2018/02/Housing-Is-Infrastructure-2018.pdf.
16. Mark Muro, Adie Tomer, Ranjitha Shivaram, and Joseph W. Kane, *Advancing Inclusion through Clean Energy Jobs, Brookings*, April 18, 2019, https://www.brookings.edu/research/advancing-inclusion-through-clean-energy-jobs/.
17. "Goal 13: Climate Action, Goal of the Month," United Nations, May 2019, https://www.un.org/sustainabledevelopment/blog/2019/05/climate-justice/.

Conclusion

1. Kimberly Vermeer and Walker Wells, *Blueprint for Greening Affordable Housing, Revised Edition* (Washington, DC: Island Press, 2020).
2. John Lewis, remarks to the 2017 Network for Energy, Water, and Health in Affordable Buildings (NEWHAB) meeting, YouTube video, May 23, 2017, 1:10, https://www.youtube.com/watch?v=3001OsLQr1M&feature=emb_rel_end.

About the Author

Dana Bourland is committed to solving our housing and climate crises in ways that advance racial, economic, and environmental justice. Over her career she has worked at the intersection of systems related to health, poverty, and the environment.

Most recently Dana led the creation of the environment program at the JPB Foundation, which is one of the largest private foundations in the United States. Formerly Dana was vice president of Green Initiatives for Enterprise Community Partners, a national affordable housing intermediary,

Photo credit: Kristin Hoebermann

where she developed and oversaw all aspects of the award-winning Green Communities program, including the creation of the Green Communities Criteria and Enterprise's Multifamily Retrofit Program.

A returned Peace Corps volunteer, Dana Bourland, AICP, is a graduate of Harvard University's Advanced Management Development Program in Real Estate and holds a master of planning degree from the Humphrey School of Public Affairs at the University of Minnesota. Dana was named one of *Fast Company* magazine's Most Influential Women Activists in Technology; she was named one of Affordable Housing Finance's Young Leaders and is featured in and has contributed to numerous publications and conferences.

Dana lives in New York City, is a devoted traveler and an Ironman finisher, and can often be found creating something using paint and a brush or a mound of clay.

Index

Page numbers in *italics* refer to figures.